PERMANENT
PRESENT TENSE

PERMANENT PRESENT TENSE

*The Unforgettable Life
of the Amnesic Patient, H.M.*

SUZANNE CORKIN

BASIC BOOKS

A Member of the Perseus Books Group

New York

Book Design by Cynthia Young
Set in Adobe Garamond Pro

Library of Congress Cataloging-in-Publication Data

Corkin, Suzanne.
    Permanent present tense : the unforgettable life of the amnesic patient, H.M. /
Suzanne Corkin.
        pages   cm
    Includes bibliographical references and index.
    ISBN 978-0-465-03159-7 (hardback)—ISBN (invalid) 978-0-465-03349-2
(e-book) 1. H. M., 1926–2008. 2. Amnesiacs—Biography. 3. Epilepsy —
Surgery — United States — History. I. Title.
RC394.A5C58 2013
616.85'2320092—dc23
[B]   2013002391

First Edition

10 9 8 7 6 5 4 3 2 1

In Memory of Henry Gustave Molaison
February 26, 1926–December 2, 2008

Contents

List of Insert Figures

9. **Mirror tracing task.**

10. **Rotary pursuit task.**

11. **Bimanual tracking task.**

12. **Coordinated tapping task.**

13. **Sequence-learning task.**

14. **Reaching task.**

15. **Basal ganglia.** Image by John Henkel, FDA staff writer, via Wikimedia Commons.

16. **Eyeblink conditioning task.** Courtesy of Diana Woodruff-Pak.

17. **Gollin Incomplete Figures Test.** B. Milner et al., "Further Analysis of the Hippocampal Amnesic Syndrome: 14-Year Follow-up Study of H.M," *Neuropsychologia* 6 (1968): 215–34.

18. **Pattern priming.** J. D. Gabrieli et al., "Intact Priming of Patterns Despite Impaired Memory," *Neuropsychologia* 28 (1990): 417–27.

19a. **William Beecher Scoville.**

19b. **Brenda Milner.** Photograph courtesy of Neuro Media Services

20. **Five-year-old Henry.**

21. **Henry the animal lover.**

22. **Henry's high school graduation photo, 1947.**

23. **Henry, 1958.**

24. **Henry with his parents.**

25. **Henry, 1975.**

26. **Henry ready for testing at MIT.** Photograph by Jenni Ogden, 1986.

27. **Henry at Bickford Health Care Center.**

28. **Henry's drawing of the floor plan of his home.** S. Corkin, "What's New with the Amnesic Patient H.M.?" *Nature Reviews Neuroscience* 3 (2002): 153–160.

29. **Henry's spoon.**

Prologue

The Man Behind the Initials

Henry Molaison and I sat opposite each other, a microphone on the narrow table between us. Parked next to him was his walker, and the white basket attached to the front held a book of crossword puzzles; he always kept one close by. Henry wore his usual attire—pants with an elastic waist, a sport shirt, white socks, and sensible black shoes. His large face, partly covered by thick glasses, wore a pleasant, attentive expression.

"How are you feeling today?" I asked him.

"I feel okay," Henry replied.

"That's good. You look great."

"Well, thank you."

"I understand that you have a little trouble remembering things."

"Yes, I do. I do have—well, a lot of trouble remembering things, you know. And one thing I found out that I fool around with a lot is crossword puzzles. And, well, it helps me in a way."

Henry and I talked a bit about his crossword puzzles, a frequent topic of conversation. Then I asked, "How long have you had trouble remembering things?"

"That, I don't know myself. I can't tell you because I don't remember."

"Well, do you think it's days or weeks? Months? Years?"

"Well, see, I can't put it in exactly on a day, week, or month, or year basis."

"But do you think it's been more than a year that you've had this problem?"

"I think it's about that. One year or more. Because I believe I had an—this is just a thought that I'm having myself—that, well, I possibly have had an operation or something."

Our conversation took place in May 1992, nearly forty years after Henry's capacity to form long-term memories disappeared as a result of a risky surgical intervention. In 1953, he underwent a bilateral medial temporal-lobe resection, an experimental brain operation intended to alleviate the severe epilepsy he had faced since childhood. Since his first seizure in 1936, his condition had grown worse, making it increasingly difficult for him to participate in normal activities. The operation did control his seizures, but with an unanticipated and devastating consequence—an extreme amnesia that robbed Henry of the ability to form new memories and, in doing so, determined the course of the rest of his life.[1]

Amnesia is the inability to establish lasting memories that are later available for conscious retrieval. The word's origin is in the Greek *amnesia*, meaning forgetfulness or loss of memory, but the deficit goes beyond forgetting. Amnesic patients such as Henry are stripped of their ability to turn their immediately present experiences into lasting memories. This condition, which may be permanent or temporary, typically stems from an insult to the brain, such as encephalitis, stroke, or head injury. Amnesia can also arise from a rare psychiatric disorder, psychogenic amnesia, which has no identified neurological cause. In Henry's case, the amnesia resulted from the surgical removal of pieces of his brain, and it was permanent.

Henry was a young man of twenty-seven when he had the operation. Now sixty-six, he relied on a walker to prevent falls. But to him, only a short time had passed. In the decades after his operation, he lived in a permanent present tense: he could no longer remember the faces of people he met, places he visited, or moments he lived through. His experiences slipped out of his consciousness seconds after they happened. My conversations with Henry vanished from his mind immediately.

"What do you do during a typical day?"

"See, that's tough—what I don't . . . I don't remember things."

"Do you know what you did yesterday?"

"No, I don't."

"How about this morning?"

"I don't even remember that."

"Could you tell me what you had for lunch today?"

"I don't know, tell you the truth. I'm not—"

"What do you think you'll do tomorrow?"

"Whatever's beneficial," he said in his friendly, direct way.

"Good answer," I said. "Have we ever met before, you and I?"

"Yes, I think we have."

"Where?"

"Well, in high school,"

"In high school?"

"Yes."

"What high school?"

"In East Hartford."

"Have we ever met any place besides high school?"

Henry paused. "Tell you the truth, I can't—no. I don't think so."

At the time of our interview, I had been working with Henry for thirty years. I first met him in 1962, when I was a graduate student. We had not met in high school, as Henry firmly believed, but by pure coincidence our lives had overlapped. I grew up in Connecticut, near Hartford, a few miles from the houses where Henry lived. When I was seven, I became close friends with a girl who lived across the street from my family. I remember her father zooming up our street in his fire-engine-red Jaguar, and on weekends, dressed in mechanic's overalls, tinkering with the car's machinery underneath.

My friend's father was a neurosurgeon. As a child, I had no idea what a neurosurgeon did. Years later, when I was a graduate student in the Department of Psychology at McGill University, this man reentered my life. While reading articles on memory in medical journals, I came across a report by a doctor who had performed a brain operation to cure a young man's intractable epilepsy. The operation caused the patient to lose his capacity to establish new memories. The doctor who coauthored the article was my friend's father, William Beecher Scoville. The patient was Henry.

This childhood connection to Henry's neurosurgeon made reading about the "amnesic patient, H.M." more compelling. Later, when I joined Brenda Milner's laboratory at the Montreal Neurological Institute, Henry's case fell into my lap. For my PhD thesis, I was able to test him in 1962 when he came to Milner's lab for scientific study. She had been the first psychologist to test Henry after his operation, and her 1957 paper with Scoville, describing Henry's operation and its awful consequences, revolutionized the science of memory.[2]

I was trying to expand the scientific understanding of Henry's amnesia by examining his memory through his sense of touch, his somatosensory system. My initial investigation with him was focused and brief, lasting one week. After I moved to MIT, however, Henry's extraordinary value as a research participant became clear to me, and I went on to study him for the rest of his life, forty-six years. Since his death, I have dedicated my work to linking fifty-five years of rich behavioral data to what we will learn from his autopsied brain.[3]

When I first met Henry, he told me stories about his early life. I could instantly connect with the places he was talking about and feel a sense of his life history. Several generations of my family lived in the Hartford area: my mother attended Henry's high school, and my father was raised in the same neighborhood where Henry lived before and after his operation. I was born in the Hartford Hospital, the same hospital where Henry's brain surgery was performed. With all these intersections in our backgrounds and experiences, it was interesting that when I would ask him whether we had met before, he typically replied, "Yes, in high school." I can only speculate as to how Henry forged the connection between his high-school experience and me. One possibility is that I resembled someone he knew back then; another is that during his many visits to MIT for testing, he gradually built up a sense of familiarity for me and filed this representation among his high-school memories.

Henry was famous, but did not know it. His striking condition had made him the subject of scientific research and public fascination. For decades, I received requests from the media to interview and videotape him. Each time I told him how special he was, he could momentarily grasp, but not retain, what I had said.

The Canadian Broadcasting Corporation recorded our 1992 conversation for two radio programs, one devoted to memory, the other to epilepsy. A year earlier, Philip Hilts had written an article about Henry for the *New York Times*, and later made him the focus of a book, *Memory's Ghost*.[4]

Scientific papers and book chapters were written about Henry, and his case is one of the most frequently cited in the neuroscience literature. Open any introductory psychology textbook and you are likely to find somewhere in its pages a description of a patient known only as H.M., next to diagrams of the hippocampus and black-and-white MRI images. Henry's disability, a tremendous cost to him and his family, became science's gain.

During his life, the people who knew Henry kept his identity private, always referring to him by his initials. When I gave lectures about Henry's contributions to science, I always encountered intense curiosity about who he was, but his name was revealed to the world only after his death in 2008.

Over the course of decades, during which I worked with Henry, it became my mission to make sure that he is not remembered just by brief, anonymous descriptions in textbooks. Henry Molaison was much more than a collection of test scores and brain images. He was a pleasant, engaging, docile man with a keen sense of humor, who knew that he had a poor memory and accepted his fate. There was a man behind the initials, and a life behind the data. Henry often told me that he hoped that research into his condition would help others live better lives. He would have been proud to know how much his tragedy has benefited science and medicine.

This book is a tribute to Henry and his life, but it is also an exploration of the science of memory. Memory is an essential component of everything we do, but we are not consciously aware of its scope and importance. We take memory for granted. As we walk, talk, and eat, we are not aware that our behavior stems from information and skills that we previously learned and remembered. We rely constantly on our memory to get us through each moment and each day. We need memory to survive—without it, we would not know how to clothe ourselves, navigate our neighborhoods, or communicate with others. Memory enables us to revisit our experiences, to learn from the past, and even to plan what to do in the future. It provides continuity from moment to moment, morning to evening, day to day, and year to year.

Through Henry's case, we gained insights that allowed us to break memory down into many specific processes and to understand the underlying brain circuits. We now know that when we describe what we had for dinner last night, or recite a fact about European history, or type a sentence on a keyboard without looking at the keys, we are accessing different types of memory stored in our brains.

Henry helped us understand what happens when the ability to store information is missing. He retained much of the knowledge he had acquired before his operation, but in his daily life afterward, he depended heavily on the memories of those around him. His family members, and later the staff at his nursing home, remembered what Henry had eaten that day, what medications he needed to take, and whether he needed a shower. His test results, and medical reports, and the transcripts of his interviews, helped preserve information about his life that he could not retain. Of course, none of these resources could substitute for the capacities Henry had lost. For memory does more than just help us survive—it influences our quality of life and helps shape our identity.

Our identity is composed of narratives we construct based on our personal history. What happens if we can no longer hold our experiences in our brain long enough to string them together? The link between memory and identity lies at the heart of our apprehensions about aging and cognitive decline. Losing our memory to dementia seems an unimaginable misfortune, yet this is what all of adult life was like for Henry. As his present moved forward, it left no trail of memory behind it, like a hiker who leaves no footprints. How could such a person ever have a clear sense of who he was?

Those of us who knew Henry recognized a clear personality—gentle, goodhearted, and altruistic. Despite his amnesia, Henry had a sense of self. But it was skewed, weighted heavily toward his general knowledge of the world, his family, and himself before 1953. After his operation, he was able to acquire only the smallest fragments of self-knowledge.

We can describe memory in all the ways we confront it in our lives. But how do our experiences translate into mechanisms in the brain? Memory is not a single event, not a snapshot fixed in celluloid with the click of a shutter. We have learned—initially from Henry—that memory does not reside in one spot in the brain. Instead, memory engages many parts of the brain in parallel. We can think of remembering as a trip to the supermarket to buy all the ingredients for beef stew. We select the meat, vegetables, stock, and spices from different parts of the store and then combine them in a large stew pot at home. Similarly, calling up the memory of one's last birthday entails pulling information stored in different parts of the brain—the sights, sounds, smells, and tastes—and organizing these stored traces in a way that allows us to relive the experience.

A popular way to think about memory in the brain is to borrow a metaphor from computer science: memory is information that the brain processes and stores. To succeed in this undertaking, the brain needs to perform three steps: it must encode the information, turning the raw data of experience into a brain-compatible format; it must store the information for later use; and it must be able to retrieve the information from storage later.

At the time of Henry's operation, little was known about how these memory processes work in the brain. In the 1960s, the discipline we now call neuroscience barely existed. Since then, Henry's case has been essential to a series of profound scientific discoveries on the nature of memory and the specific processes by which it is achieved. One basic yet crucial lesson that Henry taught us was that it is possible to lose the ability to remember, yet remain intelligent, articulate, and perceptive. It is possible to forget a conversation that took place minutes before but still have the ability to solve challenging crossword puzzles.

The kind of long-term memory Henry lacked is now called *declarative*, because people can state openly what they learned. By contrast, Henry did possess long-term memory for motor skills, such as how to use a walker; this sort of memory is now termed *nondeclarative* because people demonstrate their knowledge through performance and cannot verbalize what they learned.[5]

As neuroscience and particularly the science of memory evolved in the latter half of the twentieth century, Henry's case remained deeply relevant to research. As new theories about memory processes and new brain-imaging tools emerged, we applied them to his case. Until his death in 2008, he patiently allowed me, and more than a hundred other scientists, to study him, greatly advancing our knowledge of how the brain remembers—or fails to remember.

Because Henry donated his brain to the Massachusetts General Hospital (Mass General) and MIT in 1992, he continues to play a role in new frontiers of science. The night he died, we scanned his brain in an MRI machine for nine hours. Later his brain was preserved, and then embedded in gelatin, frozen, and cut into 2,401 ultrathin slices from front to back. These slices have been digitized and assembled into a three-dimensional image that scientists and the public will eventually be able to view on the web, offering a new way to explore in detail the anatomy of a single, well-studied amnesic brain.

We have few examples of individual patients who completely transformed a scientific field as Henry did. His story is not just a medical curiosity; it is a testament to the impact that a single subject can have. Henry's case answered more questions about memory than the entire previous century of scientific research. Although he lived his own life in the present tense, Henry had a permanent impact on the science of memory, and on the thousands of patients who have benefited from his contributions.

One

Prelude to Tragedy

In June 1939, the Molaison family was living in Hartford, Connecticut. Just across the Bulkeley Bridge was East Hartford, a city buzzing with the thrill of aviation, since it was home to Pratt & Whitney, a world leader in the manufacture of airplane engines. Pilots offered the public "sky rides" in small planes and thirteen-year-old Henry had excitedly watched these flights from the ground. He was finally going to experience one firsthand, his junior-high graduation present.

Henry and his parents drove to Brainard Field, tucked against the Connecticut River three miles southeast of downtown Hartford. There, Gus Molaison paid $2.50 for Henry to take a short tour over the city in a single-engine Ryan airplane, similar to the *Spirit of St. Louis* that Charles Lindbergh had flown across the Atlantic twelve years before— seated alone in a wicker chair, with the barest emergency supplies to sustain him and only a 223-horsepower engine to keep him from plunging into the empty sea. The Ryan's surface was polished aluminum and the interior of the cabin was lined in matte green leather. Henry settled into the copilot's seat on the right side of the plane, and the pilot showed him the controls—the yoke that guided turns and made the plane climb or descend, and the foot pedals that directed the plane's rudder.

As the engine started, the propeller began spinning and soon seemed to disappear. The pilot pushed the throttle forward, and before long the plane lifted off the runway and soared over the airport. On that spring day, everything on the ground was green and vibrant. The pilot steered them over the city of Hartford, and Henry could see the tops of the downtown buildings—Hartford's tallest building, the Travelers Tower, and the Old State House with its gleaming gold dome.

The plane was outfitted with two sets of controls, and the pilot let Henry steer. He grasped the yoke, which could be moved forward and back to point the nose up or down, or rotated to turn the plane to either side. The pilot cautioned him to hold the yoke steady and never push it forward abruptly, which would angle the plane's nose down and cause it to dive. Henry was surprised by how well he did—the plane soared smoothly under his grip.

When it was time to land, the pilot took over but allowed Henry to keep his hand on the yoke, which was mechanically linked to the pilot's. He instructed Henry on how to plant his feet on the floor during the landing to avoid inadvertently touching the rudder pedal that could cause the plane to veer. They descended, heading toward a slight bend in the river where the airport lay. As they began landing, the pilot told Henry to keep the yoke back so that the front would not point too low, causing the plane to "nose over" and land upside down. They touched down delicately and cruised to a stop on the runway.

For young Henry, that short ride must have stirred the same feelings of adventure and possibility he had felt on hearing of Charles Lindbergh's improbable journey across the Atlantic. Henry's plane ride was one of the most exciting moments of his life. Throughout the experience, he was completely entranced by the sensations of the plane, the sight of the world from high above, and the thrill of gripping the controls. Every detail of the ride was vividly recorded in his mind.

Years later, after Henry had lost his ability to form new memories, all he had left was his past—the knowledge he had acquired up to the time of his operation. He remembered his mother and father, his school friends, the houses in which he had lived, and the vacations his family had taken. But when asked to talk about these memories, he was unable to describe a unique event, a single moment in time with all the sights, sounds, and smells that came with it. He retained the general picture of his experiences but without any specific details.

Henry's first and only plane ride was one of two exceptions. Even as an old man, he could still recall the plane's green interior, the movements of the yoke, the sight of the Travelers Tower, and the pilot's instructions as he held the plane in his grip—all with perfect clarity. In decades of questioning and interviewing Henry after his operation, this was the only lengthy episode of his life that he ever described in vivid detail. The other exception was the time, at the age of ten, when he smoked his first cigarette.

Henry was born on February 26, 1926, in Manchester, Connecticut, a town about ten miles east of Hartford, in the Manchester Memorial Hospital. Henry was an eight-pound, full-term, healthy baby. His parents brought him home to their house less than a mile away on Hollister Street.

Henry's father, Gustave Henry Molaison, known as Gus, came from Thibodaux, Louisiana. After Henry's operation, he could recall his father's family origins, and would joke, "My daddy's family came from the South and moved north, and my mother's family came from the North and moved south." A relative traced the Molaison lineage back to Limoges, France. In the 1600s, the French Cajun people moved to Nova Scotia only to be deported back to France in the mid 1700s. In the late 1700s, they migrated again, this time to Louisiana, and the Molaisons settled in Thibodaux, a small community about sixty miles southwest of New Orleans. Henry's mother, Elizabeth McEvitt Molaison, known as Lizzie, was born in Manchester, Connecticut, but her parents came from Northern Ireland, where they continued to keep close family ties.

Gus was tall and lean with dark brown hair, an attractive man despite protruding ears. Lizzie was a head shorter than Gus, had curly brown hair, and wore glasses. A distant relative remembers her as having "a very mild temperament and always wearing a smile." Gus was more gregarious, often laughing and joking with friends. Gus and Lizzie were married in 1917 at Saint Peter's Church in Hartford; he was twenty-four and she was twenty-eight. The United States declared war on Germany that year, but Gus never served in the military. He worked instead as an electrician in the Hartford area, wiring buildings like G. Fox and Company, a landmark department store on Hartford's Main Street. Lizzie stayed home like most wives of the time and learned to cook Gus's familiar southern dishes. But their lives were not totally conventional. Gus and Lizzie were adventurers; they liked to take excursions, driving to Florida, Mississippi, and Louisiana to see relatives, bringing an army-surplus tent to sleep in along the way. Lizzie collected photos and mementos from these trips.

Lizzie was thirty-seven when Henry was born; he was their only child. They raised him as a Catholic. He attended a private kindergarten in nearby East Hartford and then Lincoln Elementary School in Manchester for first and second grades. By 1931, the Molaisons moved to a single-family house with a yard on Greenlawn Street in East Hartford, the first of several moves around the Hartford area that the family would make during Henry's youth. That June, Henry and his mother traveled to Buffalo, New York, for a short vacation. On a postcard to Gus, Lizzie wrote, "All well and enjoying ourselves. From . . . " and Henry, age five, scrawled his name underneath in pencil.

During the 1930s, the Molaison family lived in a residential neighborhood adjacent to the downtown Hartford area. Henry attended Saint Peter's Elementary School, next to the church where his parents had been married. He made friends, learned to roller skate, and took banjo lessons at Drago Music House on Main Street. In 1939, Henry, age thirteen, graduated from Saint Peter's and moved on to Burr Junior High School on Wethersfield Avenue. By this time, his life had begun to change.

Henry's childhood was indistinguishable from those of many middle-class boys in the 1930s. Like most boys, Henry occasionally got into accidents, and at some point he sustained a minor head injury in a bicycle mishap. The notes from medical and family sources are contradictory regarding the details: it is not clear exactly how old he was when it happened, whether he fell off his own bicycle or was struck by a cyclist while on foot, or if he lost consciousness as a result. Importantly, we have no evidence that the accident caused any brain damage, and two pneumoencephalograms (brain X-rays) carried out in 1946 and 1953, before his operation, were normal.

Yet, when Henry began having epileptic seizures at age ten, his mother would think back to the bike accident and wonder if it had caused some grave but unseen injury to his brain. Perhaps it did, but Gus's side of the family had a history of epilepsy; two cousins and a niece were epileptic, and Lizzie recalled seeing one of them, a six-year-old girl, lying stiff and inanimate in the grass at a family gathering. Lizzie later labeled this incident a "spell." She always blamed her husband's side of the family for Henry's condition. From a researcher's point of view, the cause of Henry's epilepsy could have been his minor head injury, his genetic predisposition, or both.

At first, Henry's were petit mal seizures, also called absence seizures. Far from the dramatic convulsions that many people associate with epilepsy, when one of Henry's spells came on, he simply became mentally absent for a few seconds. He did not shake, fall down, or lose consciousness; he just tuned out for a moment. If he was carrying on a conversation, he would stop talking and appear to be daydreaming. An observer might see him sway, bend his head, and begin to breathe heavily; he often made small, repetitive scratching movements with his fingers over his arms or clothing. When the seizure was over, it was as if he was waking up; he would shake his head and mutter, "I gotta come out of this again." Sometimes he might be a little dazed, but often he would resume whatever he had been doing as if nothing had happened, although he was aware of having had a seizure. These attacks occurred daily, and Henry would often explain to onlookers that he had just had a seizure.

Although Henry's petit mal seizures persisted, they never lasted more than ninety seconds, so they did not prevent him from leading a normal life. He went on vacations with his parents and played with friends in Colt Park, a playground in his neighborhood with tennis courts, baseball fields, and a skating rink. Nor did the seizures disrupt his education in either elementary school or junior high. He went to Sunday mass and studied the Catechism in preparation for his Confirmation in the Catholic Church. Notably, his seizures did not prevent him from taking the controls of the plane during his sky ride when he was thirteen.

But a drastic change occurred on his fifteenth birthday. Henry was riding in the family car, his father driving and his mother in the backseat. They were heading back to Manchester after visiting some relatives in South Coventry, a historic village about twelve miles away. Before they reached the house, Henry had a seizure unlike any he had experienced before. His muscles contracted, he completely lost consciousness, and his body shook with convulsions. His parents drove him directly to Manchester Memorial Hospital—the hospital where he was born. Afterward, he had no recollection of this episode.

This was Henry's first grand mal seizure, also called a tonic-clonic seizure after the two physical processes that occur in succession in the body—a stiffening of the limbs followed by rhythmic convulsions. Unlike the brief absences Henry had experienced before, grand mal seizures can be frightening to witnesses and exhausting for the person experiencing them. He would lose consciousness, bite his tongue, and occasionally urinate, hit his head, and foam at the mouth. These more violent seizures occurred in addition to the frequently occurring petit mal seizures and posed a serious problem for Henry and his family.

Epilepsy and *epileptic* have the same word origin as the Greek verb *ēpilambánein*, meaning to seize or attack. Epilepsy is a disease with a long history, probably dating back to prehistoric man. The first recorded episodes date from the Mesopotamian civilization in the Middle East. A manuscript from the Akkadian Empire (2334–2154 BC) describes an epileptic attack in which the afflicted turned his head to the left, stiffened his hands and feet, frothed at the mouth, and fell unconscious. Centuries of debate pitted physicians, who believed that the cause was physical and should be treated with rational manipulations, such as diet and drugs, against magicians, wizards, and quacks, who contended that epilepsy was caused by supernatural powers and must be cured with purifications and incantations aimed at appeasing the offended entity.[1]

The medical understanding of epilepsy moved forward during the sixteenth and seventeenth centuries, when scholars began to focus on the assortment of factors that preceded epileptic attacks, such as sudden fear, excitement, stress, and head injury. This shift toward a scientific interpretation of epilepsy continued during the Enlightenment, when scholars stressed observation of epileptic patients, and experimented on both animals and humans in an effort to uncover the biological causes of epileptic attacks.[2]

The nineteenth century witnessed a major advance in the study of epilepsy, as medics began to draw a distinction between epileptic patients and those considered "insane." In France, clinicians introduced the terms *grand mal, petit mal,* and *absence seizures,* and provided a detailed clinical description of each, while psychiatrists became interested in patients' behavioral abnormalities, including memory disorders.

In the late nineteenth century, the work of John Hughlings Jackson, the father of British neurology, transformed the study of epilepsy. Jackson compiled the histories of numerous cases, including his own patients, those cared for by his colleagues, and reports in the medical literature. He extracted minute details from these medical records and, based on that rich information, proposed the novel idea that seizures start in a single brain area and progress in an orderly way to other areas. This striking seizure pattern came to be known as Jacksonian epilepsy, and the initial forays into surgical treatment focused on patients whose abnormality was confined to one discrete brain area.[3]

At Jackson's suggestion, the pioneering London neurosurgeon Victor Horsley performed the first surgery for epilepsy in three patients, publishing two cases in 1886 and the third in 1909. All three patients experienced attacks in which one arm would suddenly jerk violently. During the operations, Horsley stimulated the patient's exposed brain to identify the area where the affected arm was represented. He then removed that area to arrest the spasms. In 1909, Fedor Krause, a German neurosurgeon, authored a more detailed description of epilepsy surgery. A critical part of Krause's operative strategy was his focus on stimulating the cortex electrically to map out motor, sensory, and speech areas in the human brain. These pioneering successes provided the initial validation of Jackson's insight that focal epilepsy was caused by an area of cortical irritation, and suggested that surgical treatment was safe and effective.[4]

In 1908, in the United States, Harvey Cushing at the Johns Hopkins Hospital conducted cortical localization studies in more than fifty operations for epilepsy, greatly advancing knowledge on where different functions are located in the human brain. These stimulation studies allowed surgeons to link specific behavioral abnormalities in their patients to specific areas in the cortex, an important prerequisite for epilepsy surgery. The quest to localize specific motor, sensory, and cognitive processes to identifiable brain circuits continues today in thousands of laboratories.

In the 1920s, Otfrid Foerster in Breslau, Germany, operated on patients who had brain tumors or epilepsy resulting from head injuries sustained in World War I. Foerster conducted these operations under local anesthesia, using electrical stimulation to reproduce the patients' seizures and then extirpating the offending brain area to achieve seizure control.

Foerster was mentor to Wilder Penfield, founder and head of the Montreal Neurological Institute. After a six-month visit to Foerster's hospital in 1928, Penfield returned to Montreal to expand his studies of cortical stimulation and mapping, allowing him to pinpoint and remove his patients' epileptic focus. Starting in 1939, he developed a surgical procedure, temporal lobectomy—removal of part of the left or right temporal lobe—which has since been used extensively to control seizures originating in that part of the brain.[5]

An important breakthrough at the Montreal Neurological Institute in the 1950s would profoundly affect Henry Molaison. Penfield and his colleague Herbert Jasper, a neurophysiologist, reviewed evidence from the surgical cases in which Penfield had carried out stimulation studies and from stimulation experiments in animals. They concluded that temporal-lobe seizures originated in the amygdala and hippocampus, structures deep in the temporal lobe. From then on, a standard left or right temporal lobectomy at Penfield's institute included the amygdala and part of the hippocampus. Henry's neurosurgeon, William Beecher Scoville, knew of Penfield's good results after removal of the amygdala and hippocampus, and would cite that evidence as justification for Henry's operation.

We now know that all epileptic seizures are behavioral manifestations of excessive electrical activity in the brain. Researchers first came to understand this signature of epilepsy from Hans Berger's momentous technical advance in the late 1920s. Berger, a German psychiatrist, dedicated his career to developing a model of brain function, the interaction between the mind and the brain. After disappointing efforts to link blood flow and temperature with behavior, he turned his attention to the electrical activity of the brain. In his early experiments, he inserted wires under a patient's scalp and made the first recordings of electrical activity from the human brain. Berger christened his new method the *electroencephalogram* (EEG), and with it identified different kinds of brain rhythms, some fast and some slow. After a series of technical improvements, including the introduction of noninvasive scalp electrodes, Berger succeeded in recording abnormal electrical activity in several brain disorders, including epilepsy, dementia, and brain tumors. This new window on the human brain changed the practice of neurology, giving researchers clues to the brain's underlying biology.[6]

Word of Berger's remarkable discovery reached Harvard Medical School in 1934, where it inspired a research project to study the brain's electrical activity in epileptic patients. In 1935, technician and MIT graduate Albert Grass built three EEG machines, thereby founding the pioneering Grass Instrument Company. Grass, in collaboration with neurologists William Gordon Lennox and Frederic Gibbs, recorded EEGs on paper from individuals with petit mal epilepsy. The recordings showed a characteristic pattern of brainwaves in these subjects, and later studies in other patients with grand mal seizures revealed a different distinctive pattern.

The amazing new tool, EEG, allowed doctors to identify the nature of the seizure and its location in the brain, a major advance in diagnosis and treatment. In the early days of epilepsy surgery, surgeons relied on patients' seizure patterns to identify the brain area where the seizures originated, and then removed the dysfunctional tissue. Sometimes, however, when surgeons exposed the brain in the operating room, the area in question turned out to be normal and no tissue was removed. EEG greatly improved the preoperative evaluation of epilepsy patients and also provided a means of monitoring electrical activity during operations. In the late 1930s and 40s, Herbert Jasper's laboratory devised methods to record EEG patterns and to localize seizure activity during the operation from a patient's cortex and deeper structures. Scoville and his colleagues used similar physiological recording methods during Henry's operation in an attempt to find the origin of his seizures, but to no avail. The availability of EEG machines to record seizure activity laid the groundwork for treatment with antiepileptic drugs, designed to correct the brain dysfunction evident in the EEG tracing and to prevent seizures from occurring. The use of drugs to treat epilepsy dates back at least to the fourth century BC, when practitioners administered a variety of ludicrous remedies. These treatments, some based on magical beliefs and others on observation, included camel's hair, seal's bile and stomach lining, crocodile feces, hare heart and genitals, sea-tortoise blood, and amulets made from such items as peony root. Although now considered superstitious, these remedies were said to be effective in many cases. The introduction of experimentally based anticonvulsant therapy came with the appearance of the drugs Luminal (phenobarbital) in 1912 and Dilantin (phenytoin) in 1938. In most patients, these drugs controlled seizures effectively, and became the backbone of epilepsy treatment. By Henry's time, several other anticonvulsant medications had been added to the pharmaceutical armamentarium. These drugs could lessen the severity or frequency of seizures, but often had undesirable side effects, including drowsiness, nausea, loss of appetite, headache, irritability, fatigue, and constipation.[7]

By the early 1950s, epilepsy treatment had advanced on three fronts: seizure localization, drug remedies, and surgery. Most patients achieved seizure control with personalized medication regimes. Those who required surgical intervention enjoyed satisfactory outcomes after the surgeon removed the specific area in the cortex where the seizures originated. The removals varied in extent, often restricted to part of the frontal, temporal, or parietal lobe on one side of the brain, but occasionally they included the entire cortex on the left or right. In neurosurgical centers worldwide, researchers conducted EEG studies and cognitive testing before and after operations to document treatment efficacy and to guide new approaches.[8]

At school, Henry's epilepsy kept him from fitting in. He enrolled in Willimantic High School, but dropped out for several years when he could no longer bear the teasing of other boys. In 1943, at seventeen, Henry enrolled as a freshman in East Hartford High. He was tall and quiet, with thick glasses, and he kept to himself; other than a brief stint in the Science Club, he never took part in any extracurricular activities. Few of his high-school classmates knew him; those who did commented that he was very polite.

Henry's embarrassment over his epilepsy may have kept him from being more active in school—participating would increase the likelihood that he would have seizures in front of his classmates. We can only speculate how different Henry would have been without his epilepsy, and how much of his withdrawal came from natural shyness rather than embarrassment about his disease. At that time, social attitudes toward epilepsy were still fueled by fear and misinformation, and Henry was singled out for his condition. A teacher once pulled aside one of Henry's male classmates and told him, "You're big and strong. We have a problem here: one of your classmates, Henry, has epilepsy. If he takes a fit, I want you to hold him down while I get the nurse." The student was fortunately never called into action.

An East Hartford High classmate, Lucille Taylor Blasko, remembers that the first time she noticed Henry was when she saw him on the floor of the hallway in high school, shaking and writhing. From a distance, it looked to her as if he was overcome with laughter. The next day, the school superintendent called a school-wide student assembly in the auditorium to explain Henry's situation. Although his intention was to educate the students, this assembly also singled Henry out and further publicized his condition.

Two of Henry's neighborhood friends, Jack Quinlan and Duncan Johnson, went to serve in World War II while Henry was still in school. He exchanged letters with them, and their colorful prose offers glimpses into Henry's social life. Henry was undoubtedly interested in women and went out on dates. In 1946, he seems to have admitted to Quinlan that he had a crush on an older woman. Quinlan wrote back from Chefoo, China, apparently in response to Henry's revelation: "Mi Amigo! It grieves me to find you are, beyond a doubt, a psicopathic [*sic*] case. Dames twenty-eight years old are too smart for guys like you especially the sweet married ones."[9]

Henry seems to have enjoyed other simple pleasures as well. At home, he liked listening to radio programs: he was a fan of Roy Rogers, Dale Evans, and Gabby Hayes and the family sitcom *The Adventures of Ozzie and Harriet*. He would play records on an upright Victrola, sometimes listening to popular songs with friends. Henry loved the sweetly harmonizing trio the McGuire Sisters, as well as the big bands of the 1930s and '40s, and popular hits—"My Blue Heaven," "The Prisoner's Song," "Tennessee Waltz," "On Top of Old Smoky," and "Young at Heart."

Guns fascinated Henry. With the help of his father, he amassed a collection of hunting rifles and pistols, including an old flintlock pistol, a gun popular in the eighteenth and early nineteenth centuries. Henry kept these in his bedroom, and a favorite pastime was target shooting in the countryside. He was a proud member of the National Rifle Association and loved to show off the gun collection to interested friends and relatives.

In 1947, when Henry was twenty-one, he graduated from East Hartford High. According to Mrs. Molaison, the superintendent would not let him participate in the graduation ceremony, for fear that he would have a "bad spell." Instead, Henry sat with his parents, and was "all broken up about it." In 1968, he had no recollection of the event. More than sixty of his classmates signed his yearbook, a surprising number considering his relative social isolation. It is possible that during a yearbook signing session, the books were passed around from person to person, and everybody signed everybody else's. His friend Bob Murray wrote: "A fellow roommate who lightens up the gloom." Another classmate: "To a swell fellow and a perfect friend. Love & luck always, Loris." Henry chose a quote from Shakespeare's *Julius Caesar* to accompany his handsome yearbook picture: "There are no tricks in plain and simple faith."

In high school, Henry had chosen to take the Practical Course rather than the Commercial or College Course. His classes focused on job skills over pure academics and prepared him for a technical career. At sixteen, he had a summer job as an usher in a movie theater. After graduating from high school, he first worked at a junkyard outside Willimantic rewinding electric motors. Next, he worked at Ace Electric Motor Company, also in Willimantic, where he assisted the two owners. Henry was a methodical worker, and made careful notes and diagrams about his work in a small black diary. His notes included equations for calculating voltage and power in an electrical circuit, and a diagram of two resistors in parallel. His diary also contained plans for building a model railroad. Later, Henry left the electric-motor company and worked on an assembly line at the Underwood Typewriter Company in Hartford.

Henry rode to and from work with a neighbor every day. He was unable to drive himself, as he still experienced many petit mal seizures every day and intermittent grand mal seizures. The seizures made it hard for Henry to do his job, and he often missed days. He took large doses of antiepilepsy drugs, but they could not quell his attacks.

By this time, Henry was twenty-four and had come under William Beecher Scoville's care. A prominent physician, Scoville had established the Department of Neurosurgery at the Hartford Hospital in 1939 and had a teaching position at Yale University Medical School. He held a BA from Yale University and MD from the University of Pennsylvania, and before arriving at Hartford had trained in some of the nation's top medical centers—New York Cornell Hospital and Bellevue Hospital in New York City, and Mass General and Lahey Clinic in Boston—where he was mentored by some of the most prominent figures in twentieth-century neurosurgery. Bright, energetic, and ambitious, Scoville spoke with a glint of humor but often seemed reserved to his colleagues. Regarded as an independent thinker and a nonconformist, he rode motorcycles and had a passion for old cars. In 1975, he wrote, "I prefer action to thought, which is why I am a surgeon. I like to see results. I am an auto mechanic at heart and love perfection in machinery and so I chose neurosurgery."[10]

When it was clear that the available medications were not adequate to control Henry's symptoms, his family doctor, Harvey Burton Goddard, suggested that Henry and his parents consult Scoville. Henry probably had his first appointment with Scoville in 1943, when he was seventeen, and began taking Dilantin at that time, giving him some relief from the grand mal seizures.

Sometime between 1942 and 1953, Henry's parents took him to the renowned Lahey Clinic in Boston, a trip that Henry could recount after his operation. No records describing that consultation are available. He continued under Scoville's care, so the Lahey doctors probably told the Molaisons that they could not provide any treatments not already available in Hartford, and emphasized the importance of being cared for by a local doctor. Scoville admitted Henry to the Hartford Hospital on three occasions before September 1946, but medical records from those admissions were not in Scoville's office files.

On September 3, 1946, at age twenty, Henry was admitted a fourth time, and underwent a pneumoencephalogram (brain X-ray) to rule out other abnormalities, such as a brain tumor, as the cause of the seizures. This unpleasant, invasive test was the closest that doctors could come in those days to visualizing living brain tissue without opening up the skull to peer inside. A physician inserted a needle into Henry's spine, extracted some cerebrospinal fluid, and injected oxygen that traveled up his spinal canal to his brain. The doctor then took an X-ray, which revealed the location and size of the spaces in the brain where the cerebrospinal fluid normally traveled. From those pictures, the doctor could determine whether Henry's brain had shrunk due to a disease, or whether structures had shifted to one side or the other because of an abnormal growth, such as a tumor. Patients loathed this procedure because it left them with a terrible headache and nausea. Despite these side effects, Henry left the hospital two days later with good news—his pneumoencephalogram was normal, and his physical and neurological examinations showed no problems. Although the diagnostic workup excluded some possible causes of Henry's epilepsy, such as brain tumor or stroke, it fell short of revealing exactly where his seizures originated. The Hartford Hospital discharge summary dated September 1946 read, "to continue on Dilantin indefinitely." Henry still awaited the medical breakthrough that would normalize his life.

On December 22, 1952, when Henry was twenty-six, Scoville noted that Henry had had at most one seizure within the past month. He wrote that he was "on massive medication, Dilantin five times a day, phenobarbital twice a day, Tridione three times a day, and Mesantoin three times a day." Scoville ordered precautionary, monthly blood work for Henry to be sure that the drugs did not reach toxic levels, and asked a Hartford Hospital colleague, Howard Buckley Haylett, to see Henry in his office. According to Scoville's office notes, he saw Henry again three months later, in March 1953.

Henry also endured repeated EEG studies in an attempt to find the area in his brain where the seizures originated. If his doctors had found such a focus, then they might have proposed surgically removing that area with the hope of eliminating his spells. As it turned out, however, an EEG study carried out on August 17, 1953, eight days before his operation, found only scattered, slow activity. Henry actually had an attack during the recording, which was potentially helpful, but still the EEG did not point to a specific locus of abnormality. Two days later, he had another pneumoencephalogram, and it showed no abnormalities. His vision and hearing were also deemed normal. In short, the tests available in 1953 revealed no evidence of a discrete abnormality in Henry's brain.

In a repeated attempt to unmask an epileptic focus, another EEG study was carried out the day before his operation, when he was no longer taking heavy medication. The abnormal waves were still diffuse rather than in one particular area. During the two weeks before his operation, he had had two grand mal attacks and daily petit mal attacks.

Knowing that Henry's seizures had been progressing for a decade, Scoville had suggested an experimental operation that he hoped would control the attacks and improve Henry's quality of life. He saw this operation as part of a series of investigative surgeries that could further understanding of psychiatric disease and offer solutions for certain brain disorders that seemed intractable. The operation would entail removing several inches of brain tissue from structures deep in Henry's brain, first from one side and then from the other. Scoville had performed similar operations before, but only in patients with severe psychiatric disorders, mainly schizophrenia. The psychiatric results were mixed: Scoville, in consultation with hospital staff and family members, ranked each patient's psychiatric symptoms after the operation from minus one (worse) to four (markedly improved with discharge to home). One patient was rated as minus one, and two were rated as four, with the others falling in between. Cognitive testing was not carried out. Henry would be the first patient to undergo this procedure for intractable epilepsy. In 1991 when Henry was sixty-five, a caregiver heard him say that he remembered signing forms long ago but did not remember when or what they entailed. "I think they were about my surgery on my head." No record exists of the conversations that Henry had with his parents after their meetings with Scoville, but after a decade of failed treatment, everyone agreed that the operation would be Henry's best chance for relief.[11]

On Monday, August 24, 1953, he and his parents left their home on Burnside Avenue and crossed the Connecticut River from East Hartford, driving a tense five miles to the Hartford Hospital. After being admitted, Henry met with a psychologist, Liselotte Fischer, for testing. She wrote in her report: "He admits to being 'somewhat nervous' because of the impending operation, but expresses the hope that it will help him, or at least others, to have it performed. His attitude was cooperative and friendly throughout, and he expressed a pleasant type of sense of humor."[12]

Henry spent the night in the hospital. The next day, hospital staff members shaved his head and wheeled him into an operating room. Scoville's operation report read: "Finally admitted for new operation of bilateral resection of medial surface of temporal lobe, including uncus, amygdala, and hippocampal gyrus following recent temporal lobe operations done for psychomotor epilepsy."

This was a day of eager anticipation for Scoville, and of cautious optimism for the Molaison family. Scoville knew about the procedures other surgeons were using to quell their patients' seizures, and he hoped to break new ground in surgical therapy with his own technique. Henry's case was the first in this experiment. Henry and his parents looked forward to a time when they could once again live like a normal family without the unexpected intrusions from Henry's seizures. The question on everyone's mind was would the removal of brain tissue cure Henry's epilepsy? No one anticipated that he would he lose his memory, but he did, and on that day, the entire course of his life was irrevocably altered.

Two

"A Frankly Experimental Operation"

On Tuesday, August 25, 1953, William Beecher Scoville stood over the operating table and injected an anesthetic into his patient's scalp. Henry was awake, talking to the doctors and nurses; he did not need general anesthesia because the brain does not have pain sensors and would, therefore, not register any pain during the surgery. The only places that needed to be numbed were his scalp and dura, the fibrous tissue between his skull and his brain.

When the anesthetic took effect, Scoville made an incision along a wrinkle in Henry's forehead and pulled the skin back to reveal its red underside and the bone beneath it. Just above Henry's eyebrows, Scoville drilled two holes in the skull, one and a half inches in diameter and five inches apart. He removed two disks of bone from the drill sites and set them aside. The holes became doorways to Henry's brain, through which the surgeon could insert his instruments.

Before Scoville proceeded, his team performed a final EEG study, this time with electrodes placed directly on and in Henry's brain tissue. Scoville wanted to try one last time to locate the source of Henry's seizures. The electrical activity in his brain appeared as a series of squiggled lines, called traces, on the EEG paper, with each trace corresponding to a different part of his brain. If Scoville could isolate the epileptic activity to one area, the experimental surgery he had proposed would be unnecessary; he could simply remove the discrete area from which the seizures arose. But again, the EEG showed electrical activity that was diffuse and difficult to isolate, so he moved forward with the operation as planned.

Scoville was trained in, and a strong proponent of, psychosurgery. Like many of his contemporaries, he believed that surgery offered a radical but potentially transformative solution for desperate cases. The destruction of brain tissue was, at the time, considered a valid, if experimental, treatment for numerous psychiatric diseases, including schizophrenia, depression, anxiety neuroses, and obsessional states.

Scoville believed that eventually surgeons would be able to delve into the brain and, by removing or electrically stimulating a critical area, fix problems directly, without the need for psychotherapy or drugs. Even though he was about to operate on Henry for epilepsy, not a psychiatric disease, it was through Scoville's exploration of these procedures that he came to perform such an extreme operation on Henry.

When most people think about psychosurgery, they think of frontal lobotomy, disconnecting the frontal lobes from the rest of the brain. The 1975 Academy Award–winning movie *One Flew over the Cuckoo's Nest* offers one of the most vivid cultural illustrations of the procedure. Based on the Ken Kesey novel, the movie tells the story of R. P. McMurphy, a convict sent to a mental hospital for ostensibly crazy behavior. There, he rallies his fellow patients to defy the dictatorial and hated Nurse Ratched. When his plans backfire and result in a patient's suicide, he blames Ratched and attempts to choke her to death. As punishment, McMurphy is lobotomized. The operation leaves him pitifully brain damaged, evoking the sympathy of another patient who mercifully suffocates him with a pillow (see Fig.1).

A real life example of the devastation caused by lobotomy is the well-known story of Rosemary Kennedy, daughter of Joseph Kennedy and sister to John, Robert, Edward, and Eunice. Rosemary was a pretty young woman who was said to be less intelligent than her siblings. In

1941, when she was living in a convent school in Washington, DC, the nuns reported that Rosemary was moody, had emotional outbursts, and escaped from the school at night. Concerned that she was meeting men and might get into trouble, Joseph Kennedy settled on lobotomy as a remedy for his twenty-three-year-old daughter, and took her to the renowned champion of psychosurgery, Walter Freeman. Freeman's collaborator, James Watts, diagnosed Rosemary with agitated depression, thus making her a good candidate for the procedure. The outcome was devastating and horrific: Rosemary was left mentally and physically handicapped and was institutionalized for the next sixty-three years, isolated from her family.[1]

Now banned in some countries, frontal lobotomy has been discredited and is virtually obsolete. Knowing the devastating results of these operations, it is hard to understand how they ever came to be performed. From 1938 to 1954, however, proponents of lobotomy argued that the risks of the procedure were justified by the chance to rescue desperate patients, many of whom were living deplorable lives locked up in institutions. These operations sometimes allowed patients to return to their families and resume their lives at a higher level of function than before the operation.

Surely this logic guided Scoville in his recommendation to perform the procedure on Henry. The seizures were becoming more frequent, putting Henry's life at risk, and he was no longer responding satisfactorily to even massive doses of medication. To Scoville, no doubt, surgery seemed the last, best option.

Unlike a tumor or scar tissue in the brain, which a surgeon can identify and remove, psychiatric diseases do not arise from visible changes in the anatomy of the brain or obvious disease in its tissue. The rationale for performing surgery to treat a psychiatric disease, then, is that a particular circuit in the brain is not functioning properly, even if the dysfunction is not observable.

Psychosurgery became popular as scientists began to map animal and human brains. Brain-mapping experiments began in the late nineteenth century and became increasingly popular as scientists began to understand that functions of the mind are localized in the brain. The idea behind these investigations was that specific sensory, motor, and even cognitive functions, such as language, were represented in discrete, specialized brain areas. These links between the brain and behavior, demonstrated in the late nineteenth and early twentieth centuries, raised the hope that mental illness could be localized and treated surgically.

Swiss psychiatrist Gottlieb Burckhardt published the first account of psychosurgery in 1891, removing parts of the cerebral cortex—the outside layers of the brain just under the bone—in six patients who experienced hallucinations. Burckhardt's colleagues responded to his lengthy account of his operations by ostracizing him professionally, calling his procedure reckless and irresponsible.[2]

In the early 1900s, Estonian neurosurgeon Ludvig Puusepp tried a different approach. Puusepp's three patients had manic-depressive disease or seizures, which he believed were caused by psychological disturbances. Instead of removing a chunk of brain tissue as Burckhardt had done, Puusepp cut the fibers—the "telephone lines"—that connect the frontal and parietal lobes. The surgery did not abate their disease, however, and Puusepp deemed his experiment unsuccessful.[3]

In the 1930s, psychosurgery began on a grand scale. Portuguese neurologist António Egas Moniz was a pioneer in the field whose attempts to create a biological treatment for psychiatric disorders ultimately won him a Nobel Prize. Moniz drew inspiration from an unexpected source: the Comparative Psychobiology Laboratory at the Yale University School of Medicine. Researchers there conducted experiments in chimpanzees to determine the function of the brain's frontal lobes, the part of the cerebral cortex located just behind the forehead.

In one such experiment, researchers trained Becky and Lucy, normal chimpanzees with intact frontal lobes, to perform a memory test in which they watched the experimenter hide a piece of food under one of two cups. The experimenters lowered a screen between the chimp and the cups, leaving it in place for different amounts of time, ranging from seconds to minutes. When the screen was raised, the chimp was allowed to choose one of the two cups to retrieve the reward. A correct choice reflected the animal's ability to remember where the food had been hidden. Like humans, chimpanzees manifest individual differences in personality and emotionality. Unlike Lucy, Becky had a violent dislike for the entire training experience and would not cooperate; she threw temper tantrums and rolled on the floor, urinating and defecating, and had emotional outbursts when she performed the memory task incorrectly. The researchers concluded that Becky had an *experimental neurosis*, a behavioral disorder produced in the laboratory by exposing an animal to an extremely difficult cognitive task; in essence, Becky had a nervous breakdown. Lucy, on the other hand, showed no extreme reactions.[4]

Proceeding with their experiment to examine the role of the frontal lobes in complex behaviors, the researchers removed these structures in Becky and Lucy. Postoperatively, both chimps failed the memory test when the delay was longer than a few seconds, indicating that the frontal lobes were necessary to hold the location of the food in memory. Because other intelligent behaviors were preserved, researchers knew that the chimps' failure on this task was not due to a general cognitive decline. Lucy continued to be a cooperative participant as she had been preoperatively, but Becky's behavior changed completely. In an entirely unexpected turn of events, she performed the task quickly and enthusiastically, and was no longer excitable and prone to outbursts. The researchers concluded that the frontal-lobe operation had "cured" her neurosis.

This serendipitous discovery attracted Moniz's attention. He believed that Becky's case, along with other animal studies and several clinical reports, provided enough evidence to suggest that destroying frontal-lobe tissue in humans could treat emotional and behavioral disorders. Moniz speculated that the abnormal thoughts and behaviors exhibited by psychiatric patients resulted from aberrant wiring between the frontal lobes and other brain areas. He proposed that cutting these faulty connections would redirect neuronal communication into healthy circuits, thus restoring patients to their normal state.

To achieve this result, Moniz fashioned the leucotome, a new instrument that he deemed necessary for the operation. This tool consisted of a metal tube slightly longer than four inches and about three quarters of an inch wide that could be inserted into the brain through each of two small circular holes cut in the patient's skull. Moniz's neurosurgical collaborator, Almeida Lima, initially carried out all of their operations. Lima drilled the holes, lowered the leucotome to the desired spot in the brain, and then released a thin steel wire from the base of the leucotome, which could loop out as far as about two inches from the tube. To cut the connections—the white matter—under the frontal lobes, he turned the leucotome slowly for one full circle. To make a second cut, he retracted the wire slightly and rotated the leucotome again. Then, Lima pulled the wire back inside the leucotome, removed the leucotome from the brain, plugged the hole in the skull, and repeated the procedure on the other side. The maneuver resembled coring an apple, and the effects were irreversible. Moniz called his procedure *prefrontal leucotomy*.[5]

Moniz and Lima began performing prefrontal leucotomies on humans in 1935. In Moniz's first published account of the surgeries, he described twenty patients ranging in age from twenty-seven to sixty-two. Eighteen were psychotic—they experienced irrational thinking, delusions, or hallucinations—and two were diagnosed as neurotic with anxiety disorders. Moniz described the results with this first series of patients in a 1936 monograph, in which he evaluated the therapeutic effect for different psychiatric disorders separately. He discovered that the outcome differed between psychiatric groups: patients with anxiety, hypochondriasis, and melancholy showed improvement, whereas those with schizophrenia or mania were unchanged. In his monograph, Moniz included before-and-after photographs that made his subjects appear more sane postoperatively. A close look at the individual case descriptions indicates that the outcomes were indeed mixed. Seven patients were considered cured, six showed some improvement, and seven did not benefit at all.[6]

Still, encouraged by this preliminary experiment, Moniz and Lima operated on a second series of eighteen patients. Although the surgeons had no way to evaluate the extent of the brain damage in the first twenty patients, they decided that more lesions would be better, and therefore cut six cores on each side in the second series. Moniz downplayed the severity of seizures and other troubling side effects that his patients experienced in the wake of the procedures. Notably, based on his results, he concluded that disconnecting the frontal lobes from the rest of the brain did not have "serious repercussions" on the intelligence and memory of his patients. Later, after years of performing frontal leucotomies on roughly one hundred patients, Moniz, considered the inventor of psychosurgery, turned his efforts to other interests and retired in 1944.[7]

The popularity of psychosurgery surged in the wake of Moniz's results. The operation, renamed lobotomy, was performed widely in the late 1930s and 40s. This efflorescence was due largely to Moniz's protégé, a young, ambitious American neurologist, Walter Freeman. In partnership with skilled neurosurgeon James W. Watts, Freeman performed Moniz's procedure for the first time in the United States in September 1936. Postoperatively, the patient, a middle-aged woman with anxiety and depression, enjoyed symptomatic relief and was easier to care for. Over the next three years, Freeman and Watts presented the results for their growing series of cases at scientific meetings, and the procedure gradually took hold, even at such leading institutions as the Mayo Clinic, Mass General, and the Lahey Clinic.

Freeman and Watts fine-tuned their procedures, replacing Moniz's leucotome with a new model they invented to lift up the brain and gain access to the surgical targets. This leucotome's handle was inscribed with their names. They chose to enter the skull through the temples and targeted different areas in the frontal lobes, depending on the individual patient's symptoms. Some operations were more radical than others. One modification, transorbital lobotomy, was designed to damage the thalamus—a major relay station for information entering the brain—and to minimize damage to the frontal lobes. This time, Freeman entered the brain through the bone over each eye using an instrument he found in his kitchen, an ice pick. The procedure could be carried out in ten minutes, with the patient seated in a dentist's chair. Complications included black eyes, headaches, epilepsy, hemorrhages, and death. Watts did not approve of the ice-pick operation as a routine office procedure, so the long Freeman-Watts collaboration ended, leaving Freeman to push forward on his own.[8]

The number of operations Freeman performed during his career is staggering: he carried out lobotomies on more than three thousand people in twenty-three states, not only on adult psychiatric patients but also on violent criminals and schizophrenic children, one of whom was only four years old. The majority of Freeman's patients were women, the most notable being Rosemary Kennedy. In Spencer, West Virginia, he set the dubious record of operating on twenty-five women in one day. Contrary to the Hippocratic Oath, Freeman's focus was on his procedure, not his patients.[9]

Despite the large volume of patients Freeman saw, he was determined to keep in touch with them after their operations. In 1967, he purchased a Clark Cortez camper bus, which he christened "the Lobotomobile." For years, he drove back and forth across the United States demonstrating his ice pick procedure in medical settings and visiting more than six hundred patients to note their progress. In 1967, Freeman lost his operating-room privileges at Herrick Memorial Hospital in Berkeley, California, after one of his lobotomy patients died from a brain hemorrhage. According to another psychosurgeon, H. Thomas Ballantine, Freeman also lost his privileges at Georgetown and George Washington hospitals, meaning that he could no longer admit patients or treat them there, nor could he use hospital staff or facilities. But this was as far as the medical community went to stop his harmful procedures. Shockingly, at the end of his life, the University of Pennsylvania recognized Freeman as a distinguished alumnus. He died of colon cancer in 1972 at age seventy-six.[10]

Freeman certainly was not alone in his enthusiasm for lobotomy. In the wake of his modest success, hundreds of other practitioners entered the field of psychosurgery. In the four decades after Moniz's first publication, forty to fifty thousand people were lobotomized, many against their will. But the widespread application of Freeman's lobotomy techniques did not stem from physicians' belief in Moniz's theories about disrupting the twisted wiring between the frontal lobes and other brain areas. Instead, the appeal was pragmatic; physicians had few alternative treatments to offer. The history of lobotomy is marked by optimism and a lack of skepticism on the part of both the surgeons and families of the recipients. Thousands of patients from all walks of life were operated on, often with flimsy justifications and with little evaluation and documentation of the therapeutic benefit and side effects. Women were twice as likely to be lobotomized as men.[11]

Part of the problem in the psychosurgery movement was that Moniz, Freeman, and other surgeons reported their own results, with little or no external verification. They were, of course, inclined to view their surgeries as successes and to downplay the negative outcomes. The proper evaluation of any brain operation requires, at a minimum, that the patients' cognitive abilities be tested before and after surgery to determine whether their capacities are affected by the insult to the brain. Ideally, patients should be tested by an independent psychologist who has no vested interest in the outcome, and in a way that allows for the patients' psychiatric and cognitive functioning to be quantified using standardized tests. The tests should be used to track how a patient's disease progresses over time— for better or for worse.

During the heyday of psychosurgery, few patients were given this kind of scientific scrutiny. More often, the patients' physicians, with some input from the patients' families, judged the success or failure of an operation based on subjective observations. For many of the families, any signs that the patient's behavior had improved may have been so welcome that other side effects, such as loss of memory or cognitive skills, were overlooked or accepted as a trade-off for improvement. Although these assessments were far from rigorous, the success stories were accepted—sometimes exalted—by the medical community, published in scientific journals, and heralded by the media.

Nevertheless, by the late 1950s it became clear that lobotomies were hazardous. The most tragic consequences included death, suicide, seizures, and dementia. Freeman himself recognized a *lobotomy syndrome* that could arise from the operation, with symptoms including loss of creativity, inability to react appropriately to environmental cues, bedwetting, sluggishness, and epileptic convulsions resulting from scar tissue that formed in the brain after the operation. The number of lobotomies gradually dwindled due to growing concern in the medical and scientific communities.[12]

In the later twentieth century, newly synthesized antipsychotic medications such as chlorpromazine, antidepressant medications such as imipramine, and psychotherapy began to replace psychosurgery as a form of treatment. In the 1970s, the National Commission for the protection of Human Subjects of Biomedical and Behavioral Research gathered and examined data on the effects of psychosurgery, and concluded that psychosurgical procedures should not be completely prohibited but can be performed only under certain circumstances in which the patients' rights and safety are protected. Psychosurgeons, once the stars of academic psychiatry, eventually became outsiders in the field.[13]

At the time of Henry's operation, psychosurgery was still in vogue. Still, many neurosurgeons, aware that symptoms remained after frontal lobotomy, had begun hunting for psychosurgical variants, seeking areas outside the frontal lobes that supported mechanisms underlying mental breakdown and recovery. Many researchers set their sights lower and deeper in the brain. Frontal lobotomy entailed cutting connections beneath the frontal lobes willy-nilly, whereas new procedures targeted limited brain areas.

Scoville was among the neurosurgeons developing these alternative procedures. Although he had performed frontal lobotomies in forty-three psychotic patients during the 1940s, Scoville suspected that the frontal lobes were not the seat of psychosis or the best target for curing it. He believed instead that the positive results reported for psychotics who underwent frontal lobotomies were due to the patients' reduced anxiety rather than a true change in their psychoses. Scoville, therefore, turned his focus to the inner part of the temporal lobes, which seemed to hold greater hope for curing patients. He became interested in this part of the limbic system, a set of structures underneath each cortical hemisphere, because it was believed to be the seat of emotion in the brain.

Scoville set out, as he put it, on a "project of direct surgical attack" on this part of the brain, devising a new surgical technique he called *medial temporal lobotomy*. In 1949, he began performing lobotomies focused on the limbic system. He carried out several versions of his operation, typically in female patients who were confined to state hospitals in Connecticut. Most of them were severely disturbed schizophrenics, but two patients were described as mentally deficient with psychosis and epilepsy. Scoville's procedure was intended to address the psychosis only; psychosis and epilepsy are distinct maladies, caused by different abnormalities of the brain, so it was just a coincidence that these two women suffered both. After Scoville operated on the women, their epileptic seizures became less frequent and severe; one woman showed a slight reduction in her psychiatric symptoms, and the other showed marked benefit. The seizure relief was a serendipitous finding that prompted Scoville to investigate whether temporal lobe surgery could be a treatment for epilepsy. In 1953, he published the results of the operation on the two women (and seventeen others) and operated on Henry the same year.[14]

Scoville was not alone in seeing a connection between the temporal lobes and epilepsy. Earlier studies had found that electrically stimulating temporal-lobe structures could provoke epilepsy-like symptoms in animals; the same was true when epileptic patients undergoing brain operations received electrical stimulation in these regions. In the early fifties, Wilder Penfield, an eminent neurosurgeon at the Montreal Neurological Institute, began performing operations removing tissue from the left or right temporal lobe of patients afflicted with seizures.[15]

In this context, Scoville recommended medial-temporal lobotomy for Henry. Because of the severity of Henry's epilepsy and the inability to control it, even with high levels of medication, Scoville thought he would be a good candidate for what he later called a "frankly experimental operation." He hoped that by removing a significant portion of the medial temporal lobes, he would finally be able to keep Henry's seizures at bay.[16]

Viewing the brain from the side, we see how the bulge of the frontal lobes, which fills the space behind the forehead, curves down and meets a smaller bulge lower in the brain. Scoville aimed for the inner part of this lower bulge, the temporal lobes. Gaining access through one of the holes drilled in Henry's skull, Scoville made a cut in the dura. He exposed the shiny and convoluted surface of the brain, crossed with bright red blood vessels. The brain pulsed lightly, in time with Henry's breath and heartbeat. Scoville's entry was near the optic chiasm—the area where nerve bundles running from each eye cross one another and travel to the opposite side of the brain. He inserted a long, thin brain spatula underneath one frontal lobe, lifted it up, and moved aside the large blood vessels that wrapped around the brain's surface. An assistant handed him a suction device to remove any excess blood or cerebral spinal fluid, and an electrical device to cauterize any leaking blood vessels. As he raised the frontal lobe from the lower part of the brain and spinal fluid leaked out, the brain sank down in the skull, giving Scoville more room to work. He could now see the uncus, the front part of the hippocampus. The *uncus*, meaning "hook," resembles a fist at the end of a bent wrist. Scoville had previously found that delivering even weak electrical stimulation to this structure in conscious patients caused seizures, providing a rationale for removing it to treat epilepsy (see Fig. 2a and Fig. 2b).[17]

To perform the resection, Scoville used a technique called aspiration, in which he guided a small instrument through the hole in Henry's bone and into the medial temporal-lobe region. He then applied fine suction, and with that simple action, pieces of Henry's brain were sucked into the device bit by bit. Scoville extracted the uncus, the front half of the hippocampus, and some neighboring cortex, including the entorhinal cortex. He also removed most of the amygdala, which hugs the hippocampus and is critical for expressing and feeling emotions. Having finished his work on one side of Henry's brain, Scoville repeated the procedure on the other side.[18]

The holes in Henry's skull allowed Scoville to see what he was doing, but it was still impossible to know exactly how much tissue he had extracted. Later MRI studies showed that he had overestimated the extent of the removal—he believed it was eight centimeters of tissue on each side, but the actual area missing from Henry's brain was slightly more than half that.[19]

In the course of the operation, Scoville removed the inner part of the temporal pole; most of the amygdaloid complex; the hippocampal complex, except for about two centimeters at the back; and the parahippocampal gyrus—entorhinal, perirhinal, and parahippocampal cortices—except for the back two centimeters. The brain has a left hippocampus and a right hippocampus, located above each ear deep in the temporal lobes. Pathways that cross the middle of the brain from left to right and right to left interconnect the two hippocampi. Because of Henry's case, we now know that damage to the hippocampus on both sides of the brain causes amnesia, but in 1953, scientists did not understand that the capacity for memory formation was localized to this particular area. This lack of evidence led to Henry's tragedy, and studies of his condition filled this gap in knowledge.

Before the 1930s, anatomists believed that the main function of the hippocampus was to support the sense of smell, and no one knew that a memory network occupied this structure. But scientists had written about the role of medial temporal-lobe structures in emotion. James Papez's 1937 paper "A Proposed Mechanism of Emotion" described what came to be called the Papez circuit: a ring of structures, including the hippocampus, that are anatomically connected and provide a mechanism for feeling and expressing emotion. In 1952, Paul MacLean introduced the concept of the limbic system, which included the amygdala, calling it the emotional brain. Scoville and his colleagues must have known about the central role of medial temporal-lobe structures in emotion when they carried out their medial temporal lobotomies.[20]

For Henry, the effect of removing the front half of the hippocampus was the same as if Scoville had sucked out the entire structure. The remaining two centimeters—roughly three-quarters of an inch—were deprived of input from the outside world and therefore nonfunctional. The *major* route by which information reaches the hippocampus is via pathways in the entorhinal cortex, which Scoville also removed. Thus, new information from vision, hearing, touch, and smell would not be able to reach the residual hippocampus.

Throughout the operation, an anesthesiologist carefully monitored Henry's condition. In such procedures, brain surgeons worry about damaging critical functions, such as movement and language. By asking Henry to squeeze his hand, the anesthesiologist could test both Henry's ability to understand language and his capacity to move. Although he remained conscious during the surgery, Henry was likely given sedatives to keep him from becoming restless.

When Scoville finished the removal, the anesthesiologist gave Henry a general anesthetic so he would not feel anything as Scoville completed the procedure. He stitched together the cut in the brain's outer membrane, replaced the disks of bone in Henry's skull, and sewed his scalp back together.

After the operation, Henry was taken to a recovery room, where doctors and nurses watched him closely to make sure that no life-threatening problems, such as hemorrhage, arose. Nurses checked his vital signs at fifteen-minute intervals until he was awake and clearly out of danger. They then took him back to his hospital room, where his parents were able to visit him.

In the days that followed, Henry was drowsy but otherwise seemed to make a good physical recovery from the ordeal. It soon became clear, however, that something was terribly wrong. Patients recovering from brain surgery often experience a period of confusion, but Henry's condition went far beyond that. He did not recognize the caregivers who came to his room every day or recall the conversations he had had with them, and he could not remember the day-to-day routines of the hospital. When Henry could not find his way to the bathroom despite having been there several times before, Elizabeth Molaison began to realize that something tragic had happened.

Under questioning by his family and hospital staff, Henry could recall some small events just before the time of his operation, but he did not seem to recall anything of his time in the hospital. He could not remember the death of his uncle three years prior or other momentous events in his life. By the time he left the hospital, two and a half weeks after his operation, it was clear that Henry suffered from severe memory impairment—amnesia.[21]

The operation, however, did in fact accomplish what Scoville had hoped. Henry's seizures were dramatically curtailed, but this benefit came at a devastating cost. Elizabeth and Gus, who always had to take care of Henry because of his seizures, now found themselves with a son who could not remember what day it was, what he had eaten for breakfast, or what they had said just minutes before. For the rest of his life, Henry would be trapped in a permanent present tense.

Three

Penfield and Milner

After his operation, Henry might have continued to live a difficult but private life under the devoted care of his parents. But his case soon attracted the attention of the scientific community, hungry for knowledge about the human brain. From his tragedy, we learned that our brains are capable of carrying out many different computations related to memory as it is formed, consolidated, and retrieved in numerous, specialized brain circuits.

Henry was not the first person to develop a severe, long-term memory impairment following a brain operation to relieve epilepsy. Around the same time, two other men, F.C. and P.B., suffered a similar plight. Both men were amnesic immediately after operations performed by Wilder Penfield, the founder and director of the Montreal Neurological Institute at McGill University. Penfield had removed part of the left temporal lobe in each patient to alleviate epileptic seizures.[1]

Penfield, along with then McGill graduate student Brenda Milner, studied both F.C. and P.B. extensively. Milner later studied Henry as well. Their research was part of a growing movement among scientists to link complex mental abilities, such as memory and cognition, to specific anatomical structures in the brain. These three remarkable cases—F.C.,

P.B., and H.M.—provided a quantum leap in neuroscience and formed the basis of modern memory research.

Henry's story is inextricably linked to the extraordinary life of Penfield and his institute. Penfield was born in Spokane, Washington, in 1891. His father and grandfather were physicians, and he followed in their path. After attending a private high school for boys, he studied at Princeton University and later began medical school at the College of Physicians and Surgeons in New York. But Penfield's plans changed after six weeks. His academic achievement, athletic prowess, and social success won him a Rhodes Scholarship, and in 1914, at age twenty-four, he entered Merton College at Oxford University in England.

My brief account of Penfield's life draws heavily from his autobiography. He studied both science and medicine, and this dual training established his lifelong passion for bridging the two disciplines. From the outset of his studies, Penfield worked under giants in these fields. During his first two years at Oxford, his mentors were Sir Charles Scott Sherrington, winner of the Nobel Prize for Physiology or Medicine in 1932 for his discoveries on the function of neurons, and Sir William Osler, architect of bedside teaching and the medical residency system.[2]

After completing his fellowship at Oxford, Penfield returned to the United States and finished his last year of medical school at Johns Hopkins University. Following a surgical internship in Boston at the Peter Bent Brigham Hospital, under the famous neurosurgeon Harvey Williams Cushing, Penfield returned to England for two years of graduate study, pursuing neurophysiology in Oxford and neurology in London at the renowned National Hospital at Queen Square.

In 1921, Penfield returned home, and at age thirty, with unrivaled schooling, he accepted a position at the Presbyterian Hospital in New York for training in neurological surgery. There he made his initial venture into neurosurgery. His first patient was a man with a brain abscess, a mass filled with pus. The second was a woman with a brain tumor. Both patients arrived at the hospital in comas, and despite Penfield's heroic attempts in the operating room to save them, both died. Although depressed by these failures, Penfield believed that the practice of brain surgery would make enormous leaps during his lifetime.

The focus of Penfield's teaching and research was the examination of tissue he had removed from patients' brains during surgery. He hoped that through his microscope, he would see something to provide clues about the cause of epilepsy. His results were disappointing, however, because his methods could not capture sufficient detail in the cells. Around that time, he had the good fortune to read an article in a Spanish journal that included drawings of brain cells, in which the different parts of each cell stood out sharply. The author of the article was Pío del Río-Hortega, a Spanish researcher at Madrid's Cajal Institute, and Penfield was eager to visit his laboratory. In 1924, he received permission from his department to go to Spain for six months to visit Río-Hortega's lab. There, Penfield worked on a fundamental problem that faced biologists: how to identify specific types of cells.

When researchers look at brain tissue through a microscope, they see a complex and mysterious array of structures. The brain has many different kinds of neurons, which are specialized for different functions. But as important as neurons are, *glial cells* far outnumber them. Glia—Greek for *glue*—provide structural support for neurons, and, in Penfield's day, were believed to be unimportant for the transmission of nerve impulses. We now know, however, that they are active partners with neurons, and that the interactions between these two cell types are likely vital to the function of synapses, the gap across which one neuron sends messages to the next.

Researchers often study neurons and glia by injecting stains that are taken up by a specific kind of cell, making it stand out from its neighbors. In Madrid, Penfield helped pioneer this technology with Río-Hortega, who had developed advanced methods for staining brain tissue as a tool for uncovering the structure of nerve cells and their connections. Under Río-Hortega's guidance, Penfield produced the first reliable stain for a kind of glia called *oligodendroglia*, and described them in a 1924 publication. Because these cells appear in response to brain disease or injury, being able to identify them gave neuropathologists a method for examining abnormal brain tissue.[3]

Penfield was particularly curious why a scar in the brain caused by birth injury or head trauma could lead to epilepsy in some patients. He had the opportunity to pursue this question four years after his visit to Spain. His fascination with epilepsy spurred another European trip, a six-month visit to Otfrid Foerster's laboratory at the University of Breslau in Germany. Foerster had been carrying out operations on epileptic patients, extracting the scarred brain tissue that was causing their seizures. This procedure was the exact one Penfield wished to perform himself, and he was eager to observe Foerster's work, step by step, in the operating room.

In 1928, Foerster invited Penfield to watch him remove from a patient's brain a scar resulting from a gunshot wound sustained sixteen years earlier. Penfield was able to take the scar tissue to a small laboratory specially equipped to apply the new Spanish staining techniques. There he discovered what he had been searching for: glia. He saw the cells he had viewed before in other injured brains, but this time could see them in greater detail, with all their complex branches clearly discernible. This exciting breakthrough was one of the highlights of Penfield's life. He was face to face with the cellular aberrations that caused the patient's seizures, a discovery fundamental to understanding how disease or injury to the brain, and scar formation during healing, caused epilepsy. Understanding the cause opened the door to finding a cure.

Penfield's findings excited Foerster, who proposed that they collaborate to publish descriptions of twelve cases of similar operations he had performed, all of which had led to improvements in his patients. Foerster wanted Penfield to examine microscopically the brain specimens he had extracted during his successful operations for epilepsy. This tissue held evidence on the cause of the patients' seizures.[4]

During the rest of his stay in Breslau, Penfield documented the presence of microscopic abnormalities in brain tissue from twelve patients who experienced excellent seizure control for up to five years after their operation. In their joint paper, Penfield and Foerster combined the positive results of the surgery with descriptions of the abnormal brain tissue, thereby linking the cause and the cure. Foerster's operation seemed a promising way to provide seizure control for desperate patients, and it gave Penfield a vital tool for the future. He now had the scientific justification he wanted for removing abnormal tissue, as Foerster had done, using only local anesthesia. With the patient conscious and cooperating during the operation, Penfield could stimulate the brain to map out motor and language areas, which he would not remove, and identify areas of abnormality to be excised. With the expectation of curing their epilepsy, Penfield would use this approach with a large group of patients. This epiphany set the stage for the rest of his career.

In 1928, Penfield took up residence in Montreal to pursue a long-held dream, establishing a specialized neurological institute at McGill University. His plan—which became his life's mission—was to build an institute near, but independent of, a general teaching hospital. He envisioned a neurological institute that would combine facilities for patients and researchers in the same building, providing a central point of research and discovery for the region. A major asset in this ambitious effort was neurosurgeon William V. Cone, Penfield's first student and a close collaborator in New York, who moved with him to Montreal. Penfield described Cone as a "brilliant operator and technician"—a scholarly man dedicated to patient care, to perfecting the discipline of surgery, and to discovering innovations in pathology. In Penfield's view, they were "fellow explorers," and working with Cone made him "twice as effective."

Penfield was a skilled collaborator, and one of his early successes was bringing neurologists from different hospitals in Quebec together for a weekly conference to exchange insights about puzzling or unusual cases. Those discussions forged a new bond between English-Canadian and French Canadian neurologists. Penfield's vision for a neurological institute extended this sort of collaboration on a more ambitious scale. But this plan could be realized only with major financial backing, and after initially being denied a grant from the Rockefeller Foundation, Penfield obtained his funding from two unlikely sources.

The first gift came from the mother of a sixteen-year-old boy who had uncontrolled major epileptic attacks from injuries likely sustained by the use of forceps during his birth. In gratitude for Penfield's taking on her son's case, the boy's mother sent him an unsolicited check for $10,000 to expand his research efforts in epilepsy. Before operating on the boy, Penfield used her money to consult several senior colleagues about the surgical options. He then performed what he termed a "frankly exploratory operation," in which he removed an artery from the left side of the boy's brain, believing it was causing the attacks. The procedure was successful and relieved the boy's seizures. Eighteen months later, the mother died of cancer and left Penfield $50,000 to pursue his mission.

The second windfall came from the father of a young man with unlocalized epileptic seizures. Penfield performed a radical operation in which he removed nerves connected to the arteries that entered the skull. Before the surgery, he informed the boy's parents that he had previously performed this operation in monkeys without ill effect. "The boy was greatly improved by the operation, if not cured," Penfield later reported. The patient's father was a member of the Board of Trustees of the Rockefeller Foundation, and after the surgery was over, he discussed Penfield's work with the new director of the Rockefeller Foundation's division of medical education, Alan Gregg.

In March 1931, Penfield met with Gregg in his lower Manhattan office, where the panoramic view from the twenty-seventh floor took in the Hudson River, East River, and Long Island Sound. In this delightful setting, the two men had a long, friendly discussion about neurology, neurosurgery, and research in Europe. Being cautious, Penfield did not mention his hopes for a neurological institute, nor did Gregg hint at Rockefeller's plans for funding. After he returned home, Penfield sent Gregg a warm invitation to visit Montreal.

Seven months later, Gregg visited Penfield at his home and astonished his host when he pulled the original application from his briefcase, placed it on the coffee table, and said, "This is exactly the sort of thing for which we are always searching at the Rockefeller Foundation. . . . I think I understand what you want to do. . . . Don't ever thank us. We thank you. You will be helping us when you do your job." Penfield received $1,232,000 from the foundation.

The Montreal Neurological Institute, L'Institut Neurologique de Montréal, opened its doors in 1934. "The Neuro" epitomized the wisdom of advancing science, teaching, and patient care under a single umbrella. Around the Institute, Penfield became known as "The Chief"; he was a skilled and innovative neurosurgeon as well as a strong leader. He further developed the approach he had witnessed in Breslau of operating on epilepsy patients while they were awake and conscious so that he could pinpoint the abnormal tissue responsible for their seizures—a technique that came to be known as the Montreal Procedure. These operations opened up new possibilities for scientific discoveries about specialization in the human brain.[5]

Brenda Milner, who became a crucial player in the development of memory science, was a graduate student in Psychology at McGill University when she began her collaboration with Penfield. Born in Manchester, England, in 1918, she studied experimental psychology during her undergraduate years at the University of Cambridge. There her mentors included a prominent experimental psychologist and memory theoretician, Sir Frederic Bartlett. Her research supervisor, Oliver Zangwill, also an experimental psychologist, pioneered the study of neurological patients and was keenly interested in memory disorders.[6]

Milner moved to Montreal in 1944 and two years later had the distinction of being a student in the first seminar at McGill University taught by Donald O. Hebb, a physiological psychologist who was highly influential in the science of learning and memory. Three years later, Milner became Hebb's graduate student. When Penfield invited Hebb to send someone from his lab to study his surgical cases, Milner jumped at the opportunity. Her mandate was to design and conduct cognitive research with epileptic patients, creating tests to assess their capabilities both before and after surgery to document its effects on the brain. Thus, one of the great partnerships in the history of science was born.

In the early 1950s, Milner and Penfield undertook a detailed study of two patients whose case histories were highly atypical. The patients, F.C.

and P.B., were noteworthy because they provided new data about the vital function of structures in the inner part of each temporal lobe—the same structures that Scoville would eventually remove from Henry's brain.[7]

Penfield operated on a great many epileptic patients to control their seizures, and F.C. and P.B. were part of that group. Before their operations, they did not present differently from the other patients. But afterward, both faced a complication that no one had anticipated: a lasting inability to record new experiences. Why this aberrant outcome?

To answer this question, Penfield needed to understand how the damage in these two patients differed from that in the other patients in his series. Both men had undergone a standard procedure: a partial left temporal lobectomy. In this operation, Penfield typically removed the cortex—the superficial layer—of the lateral temporal lobe, along with varying amounts of tissue deep in the temporal lobe: the amygdala, hippocampus, and neighboring cortex. Penfield had not done or noticed anything unusual during the operation on either man.

As soon as F.C. recovered from the operation, it was clear that he was unable to form new memories. P.B.'s case was slightly different, in that his surgery was carried out in two stages five years apart, and only after the second stage did he become amnesic. His first operation was similar to F.C.'s, but less tissue was removed: Penfield spared the hippocampus and other deep temporal-lobe structures. But after he went home, P.B. continued to have seizures, so five years later, Penfield operated again, this time removing the hippocampus and surrounding tissue. When P.B. recovered from the procedure, he too was amnesic.

Although the resultant severe memory loss pointed to the hippocampus and surrounding tissue as the culprit, Penfield could not understand why the damage in these two men was different from that of the other patients who underwent a left temporal lobectomy. In most of those cases, as in F.C. and P.B., part of the left hippocampus was included in the removal. Why then were these men amnesic when dozens of other patients who received a similar operation were not?

Penfield and Milner suspected that F.C. and P.B. might have some undetected abnormality in the corresponding area in their right hippocampal region. They reasoned that the abnormality discovered in the left temporal lobe at the time of the operation was likely due to a birth injury, which could have affected the right medial temporal lobe as well. Herbert Jasper, a renowned neurophysiologist at the Neuro, who conducted several EEG studies with F.C. and P.B, ultimately proved this hypothesis correct. In both men, Jasper found clear evidence of damage to the hippocampal region on the non-operated side of their brains. This abnormality, which was related to their epilepsy, was not obvious before surgery but was apparent in the postoperative EEG studies.[8]

In 1964, researchers obtained further clarity on P.B.'s case. After he died of a heart attack, his wife allowed Penfield to examine P.B.'s brain in the laboratory to probe the cause of her husband's amnesia. When Gordon Mathieson, a neuropathologist at the Neuro, examined the brain, he found that the right hippocampus was shrunken, with only a small number of neurons surviving. This massive destruction likely stemmed from an injury at the time of his birth.[9]

Unlike the other patients in Penfield's series who still had one normal temporal lobe, these men had two abnormal temporal lobes—the one removed during the operation, and the one that remained. It was this double loss that made them exceptional. These cases showed that the anatomical foundation for amnesia is a loss of function in both hippocampi. But if a person has damage to only one hippocampus, either the left or the right, the result is not catastrophic. Subsequent research on hundreds of patients has taught us that the hippocampus can be removed safely on one side with only minor memory impairment, as long as the other hippocampus is intact. One hippocampus by itself apparently can compensate largely for its missing twin, suggesting that the two structures share a general capability for making memories. Details of brain anatomy may explain this sharing of function. We know that the left temporal lobe is specialized to process verbal information and the right temporal lobe visual-spatial information. Anatomical bridges that cross the brain from left to right and right to left give each temporal lobe access to the specialized information from the other side. When one hippocampus is missing, the remaining one can engage multiple kinds of knowledge, both verbal and nonverbal, to support satisfactory learning and memory.[10]

Milner evaluated F.C. and P.B.'s cognitive capacities before and after their operations, using measures of their overall intelligence and memory. By comparing these two sets of test scores, she identified ways in which their cognitive function was altered, or unaltered, by the left temporal lobectomy. She could then link any deficits she uncovered to the damaged brain structures. Their amnesia, which Milner meticulously documented, was all the more striking because it occurred against a background of normal intelligence. Before their operations, F.C.'s IQ was average and P.B.'s was above average. Both men showed no change in their IQ after their operation—in other words, they were still capable, intelligent men. They could repeat strings of digits forward and backward and do simple arithmetic problems in their head, indicating that they could attend to and accurately perceive test stimuli and keep them in mind for a few seconds. Despite these intact cognitive abilities, however, they both failed to remember new information. Their long-term memory capacities were shot—and would never recover.[11]

The results for F.C. and P.B. made it clear that their memory loss was not limited to a certain kind of information, but instead encompassed all kinds of test materials, public and private events, and general knowledge. F.C. and P.B. had a *global amnesia*. As is usually the case, their immediate recall or recognition of the test stimuli and everyday events was better than after several minutes or hours had passed; the passage of time takes its toll on memory. But F.C. and P.B.'s amnesia was not complete; unlike Henry, they each retained a smidgen of long-term memory capacity to guide them through everyday life. F.C. was able to resume his job as a glove cutter, and P.B. his job as a draftsman.

In 1954, Milner presented the psychological test results for F.C. and P.B. at the annual meeting of the American Neurological Association in Chicago. Prior to the meeting, Scoville read a long abstract of Milner's talk and called Penfield to tell him about his two similar cases of amnesia, H.M. and D.C. Penfield was already interested in memory mechanisms, and Scoville's cases attracted his attention because they supported his ideas on the neural localization of memory. Accordingly, Penfield asked Milner whether she would be interested in testing Scoville's patients, and she embraced the opportunity. With this collaboration already underway at the time of the meeting, Scoville was asked to lead a formal discussion following Milner's presentation. He described his operative technique and results for thirty patients, twenty-nine of whom were schizophrenic and one who suffered from intractable seizures. All had the medial temporal-lobe structures removed, but two had received more extensive operations. Henry, the epileptic patient, was one of them.[12]

The other patient was a forty-seven-year-old physician with paranoid schizophrenia, known by his initials, D.C. Over time, D.C. had become violent and combative, and tried to kill his wife. He was institutionalized, and drastic treatments—insulin-shock therapy to induce a coma, and electroconvulsive therapy to trigger a seizure—failed to improve his situation. In a last-ditch effort to help D.C., Scoville traveled to Manteno State Hospital in Illinois in 1954 and performed his bilateral medial temporal-lobe resection, assisted by John F. Kendrick Jr., a neurosurgeon from Richmond, Virginia. This operation, in which D.C.'s hippocampus and amygdala were removed on both sides, took place approximately nine months after Henry's. Postoperatively, D.C.'s aggressive behavior disappeared, and although he still showed signs of paranoia, he became more friendly and manageable. Like Henry, he exhibited profound memory impairment and was unable to find his way to his hospital bed or recognize the staff.[13]

During his American Neurological Association talk, Scoville highlighted one striking behavioral result, a near-total loss of recent memory in two patients that was unaccompanied by personality change or intellectual decline. Although Scoville's clinical descriptions of Henry and D.C. were compelling, they lacked the rigor of a thorough, systematic investigation. It was important to probe the cognitive capacities of the two men, one by one in formal experiments. Cognitive deficits are often subtle and can be overlooked without measuring performance numerically and comparing patients' scores to the scores of healthy individuals.

Penfield arranged for Milner to review nine of the thirty patients who had received bilateral or unilateral medial temporal-lobe operations at Scoville's hand, and who were sufficiently stable to undergo testing. Henry was one of them.

The results of Milner's psychological evaluation of Scoville's patients formed the basis of Scoville and Milner's "Loss of Recent Memory after Bilateral Hippocampal Lesions," the benchmark *Journal of Neurology, Neurosurgery, and Psychiatry* paper. This often-cited article provided the scientific evidence for the pattern of memory loss with preserved intelligence Scoville had seen in his initial clinical evaluation of Henry and D.C. This paper has become a classic in neuroscience literature for several reasons: of greatest importance, it informed neurosurgeons that destroying the medial temporal-lobe structures on both sides of the brain would cause amnesia and should be avoided. The results also established, for the first time, that a distinct region of the brain—the hippocampus and its neighbors—was necessary for long-term memory formation. The Scoville and Milner article also inaugurated decades of experimental studies of Henry and other amnesic patients, and inspired animal models of amnesia that yielded a wealth of information on the biology of memory processes.[14]

Milner examined Henry for the first time in April 1955, twenty months after his operation. She gave him every cognitive test she could lay her hands on, and her findings launched a new era in the science of memory. Her formal testing showed that Henry's overall intelligence was above average and that his capacities for perception, abstract thinking, and reasoning were normal. But when she probed his ability to remember information beyond the immediate present—his long-term memory—his deficit was obvious, despite his excellent motivation and cooperation. Henry performed the same memory tests that F.C. and P.B. had taken, but his scores were even worse. When asked to recall brief stories and geometric drawings, his scores were far below average, and in some cases zero. Throughout testing, Milner was struck that once Henry switched to a new task, he could no longer recall the preceding one or recognize it when it was repeated. Any distraction immediately put him at a total loss.[15]

Henry's amnesia was more profound than that of F.C. and P.B., likely due to greater damage to his medial temporal-lobe structures, and he gradually became the yardstick against which other amnesic patients were judged. In the scientific literature, they were deemed "as bad as H.M." or "not as bad as H.M." Because his amnesia was not intertwined with a psychiatric disorder, Henry was a more straightforward case than D.C. Because his operation did not cause any cognitive deficits other than amnesia, his performance on memory tests was a pure measure of his memory capacities. Henry became the gold standard for the study of amnesia.

Scoville and Milner concluded their celebrated paper by identifying the hippocampus and adjacent hippocampal gyrus as the substrate for re-membering new information. The severity of memory loss in all ten cases was related to the size of the hippocampal removal—the larger the re-moval, the greater the memory impairment. What distinguished Henry from patients with amnesia due to other causes, such as Alzheimer disease or head injury, was that his memory impairment was amazingly specific. The purity of his disorder made him a perfect focus for the in-vestigation of memory mechanisms in the human brain.[16]

As Milner delved deeper into the study of memory loss, and Henry's in particular, Scoville moved on. He maintained an active neurosurgical practice and published more than fifty papers in medical journals, but did not continue to see Henry. I know firsthand, however, that Scoville was still interested in Henry's case. In the late 1970s, when I was visit-ing my parents, who lived across the street from him, he invited me to his house to get an update on Henry and our research with him.

In his writings and lectures, Scoville shared Henry and D.C.'s cata-strophic losses with the medical community for their scientific value. In the interest of a larger cause, he warned other neurosurgeons against damaging the hippocampal area on both sides of the brain, and they took his warnings to heart. In a 1974 lecture, he called Henry's operation "a tragic mistake." According to his wife, he "deeply regretted" what he had done to Henry. In 2010, Scoville's grandson, Luke Dittrich, wrote an article for *Esquire* magazine, in which he gave a colorful account of his grandfather's life and career.[17]

In 1961, I joined Milner's laboratory at the Neuro as a McGill University graduate student. This institution was renowned for the treatment of epilepsy patients, using the surgical procedures Penfield had developed. In Milner's lab, our research focused on these patients. Milner was especially skilled at designing tests that could be given before and after an operation to tease apart a patient's performance on different cognitive tasks—sensory perception, reasoning, memory, and problem solving—to discover any changes in brain function caused by the surgery. We communicated closely with the surgeons and knew after each procedure what part of the patient's brain had been removed and the size of the excision.[18]

In addition to conducting preoperative and postoperative testing, I had the opportunity to witness the operations on my patients' brains. From behind a glass window in the viewing gallery in the main operating amphitheater, I could look over the surgeon's shoulder at the patient's exposed brain and watch the surgeon stimulate the brain to map out landmarks before removing any tissue. To guard against damaging areas specialized for language and movement, the surgeons identified these regions by electrically stimulating the outer layers of the brain while the patients were awake. When this stimulation interrupted the patients' speech, caused a spontaneous movement, or made them think of a particular object, face, sound, or touch, the surgeons placed a small letter on the stimulated brain area. A stenographer seated next to the operating table noted the behavior associated with each letter. Photographs of the brain showed the letters and later gave clues about localization of functions in the cortex. The electrical stimulation also helped identify where the epileptic seizures originated, and that area would be excised. I could see which parts of the brain were removed, and the extent of each procedure was later spelled out in a report with photographs and the surgeon's drawing of the location and size of the removal.

This documentation was crucial for making sense of the behavioral test scores that we collected in the lab. By combining test results with the surgeons' reports, we could link any cognitive deficits in our patients with the brain areas that were lost, and their normal performance to the areas of the brain that remained intact. With this collaborative approach, Milner and her colleagues made important discoveries about the organization of the left and right cerebral hemispheres in humans, based on the actual necessity of each brain region for a particular cognitive process.[19]

My PhD thesis project studied how operations to alleviate epilepsy affected the somatosensory system—the sense of touch. To do this, I devised and constructed memory tests that required patients to rely on touch rather than vision or hearing. I tested many patients who had brain tissue removed from either the left or right frontal, temporal, or parietal lobe. I was particularly eager to test the three amnesic patients who had lesions in both hippocampi—patients about whom I had read earlier in the papers that Milner had coauthored with Penfield and Scoville: F.C., P.B., and Henry. Epilepsy surgery did not typically result in amnesia; these three cases were rare.[20]

I first met Henry in May 1962, when Milner arranged for him to visit us at the Neuro for testing. This was his first and only trip to Montreal, and it was momentous. He and his mother came by train, which is how they traveled long distances. Mrs. Molaison feared air travel, and the train was less costly. They stayed in a nearby rooming house, and every morning for a week, the two of them arrived at the Neuro and made their way to the Neurology waiting room.

During that week, my colleagues and I took turns testing Henry. Each day, I picked him up in the waiting room and guided him to my testing room, and when we finished, I escorted him back. He was a cooperative research participant, as he would be for the rest of his life, and we completed all the tasks I had planned for him. Even then, I felt privileged to work with Henry, along with F.C. and P.B.—a rare trio of amnesic patients. But in 1962, I had no idea how famous Henry would become.

At the time of his visit to Montreal, Henry was in his thirties, in the prime of his life, but completely dependent on his mother. Mrs. Molaison, a housewife and Henry's constant caretaker, was a pleasant, sweet woman. During the entire visit, she sat patiently in the dreary waiting room while researchers took her son to various testing rooms. She was terrified of the big city where people spoke French, a language she did not understand, and preferred to stay within the safe walls of the Neuro rather than explore on her own.

Henry's week at the Neuro was busy. We had prepared an extensive series of tests for him, designed to measure various facets of his memory and other cognitive functions. Although we did not know it then, the results of our studies would reveal the scope and limits of his amnesia, and in doing so would foreshadow new ways of exploring how memory is organized in the human brain. His memory loss, while having a devastating impact on his daily life, proved a priceless gain in the quest for the underpinnings of learning and memory.

Four

Thirty Seconds

From the beginning, one of the most striking aspects of Henry's memory loss was how remarkably specific it was. He forgot all of his experiences after his 1953 operation, but retained much of what he had learned before that. He knew his parents and other relatives, recalled historical facts he had learned in school, had a good vocabulary, and could perform routine daily tasks, such as brushing his teeth, shaving, and eating. Studying Henry's remaining capacities proved just as instructive as studying those he had lost. One important lesson scientists have learned from people with selective memory loss such as Henry's is that memory is not a single process but a collection of many different processes. Our brains are like hotels with eclectic arrays of guests— homes to different kinds of memory, each of which occupies its own suite of rooms.

Henry's case shed light on a longstanding controversy about whether brief memory mechanisms are distinct from lasting ones. The basic question was whether the processes that support short-term memory, which holds a limited amount of information temporarily, differ from those that support long-term memory, which hangs on to vast amounts of information for minutes, days, months, or years.

Most of us use the term *short-term memory* incorrectly. Short-term memory, as defined by memory researchers, does not refer to recalling what we did yesterday, this morning, or even twenty minutes ago. That sort of recollection is recent, long-term memory. Short-term memory is the immediate present, the information on our radar screens at this very moment; it expires within about thirty seconds or less, depending on the task. Its capacity is limited, and it fades immediately if we do not rehearse it or convert it into a form that can be retained in long-term memory. When I tell a friend my phone number, the digits will remain in her short-term store briefly, and she will quickly forget them unless she mentally processes them or writes them down. The short-term store is not a warehouse in the brain; instead, it is a series of processes that keep bits of information, such as my phone number, active for a brief period of time. Long-term memory, on the other hand, is anything we remember after just seconds have elapsed.

Was the formation of short- and long-term memory part of a single process, or instead, governed by wholly separate processes? Those who supported the dual-process theory sought convincing evidence that a particular patient was impaired on tests of long-term but not short-term memory, and that another patient was impaired on short-term memory tasks but not long-term. These two results, taken together, would indicate that the two kinds of memory were independent. Studies of patients with selective damage to their brains sharpened the debate over whether memory is a single or dual process, and Henry played a starring part in this research.

Henry's role as a research participant began in 1953, just prior to his operation. Scoville ordered a complete psychological evaluation to establish a preoperative baseline against which to measure any changes resulting from the procedure. The day before his surgery, clinical psychologist Liselotte K. Fischer sat down with Henry at the Hartford Hospital and conducted a series of tests, including an IQ test, a memory test, and several others designed to reveal his personality and psychological status. One task is a common measure of short-term memory called *digit span*, in which an examiner asks a patient to repeat a gradually increasing series of numbers. For example, if she says, "Three, six, nine, eight," the patient immediately repeats, "Three, six, nine, eight." The researcher then gives five digits, then six, then seven, eight, and so on; if the patient repeats eight digits but fails at nine, then the patient's digit span is eight. Fischer administered this test to Henry, and then asked him to repeat strings of digits in reverse order, a much harder task. If she said, "Three, six, nine," then the correct answer would be "Nine, six, three." His combined score for both tests was six—well below the normal range.

Two years after Henry's operation, when Milner gave him a similar test, his digit span had improved, putting him in the normal range. His ability to remember more digits postoperatively, however, does not mean that the surgery improved his memory. Multiple factors may have contributed to his weak preoperative performance. During testing, Fischer witnessed several petit mal seizures, which were not unexpected since Henry had been taken off his medication in preparation for surgery. In addition, he was anxious about the upcoming operation, triggering stress-related mechanisms in his brain that may have interfered with his test performance and masked his true abilities. His deficits before the big event were likely a combined result of seizures and nerves.

Over the decades when my colleagues and I studied Henry, he maintained a normal level of performance when we tested his digit span. This finding presented a sharp contrast: Henry suffered a catastrophic memory loss, yet he could briefly remember and repeat a string of digits. This suggested that Henry's short-term memory was intact; where he failed was in converting short-term memories into long-term memories. For instance, during the course of a fifteen-minute conversation, he would tell me three times the same story about the Molaison family's origins without knowing that he was repeating himself. The information could be collected in the hotel lobby of Henry's brain, but it could not check into the rooms.

William James, a brilliant psychologist and philosopher, was the first to make a distinction between two kinds of memory. In 1890, he produced an often quoted, two-volume tour de force, *The Principles of Psychology*, in which he described primary and secondary memory. *Primary memory*, he said, makes us aware of "the just past." The content of primary memory has not yet had a chance to leave consciousness; primary memory covers such a short span of time that it is considered "right now." Reading these sentences, we are simply carrying all the words in our minds at the present moment, not actively dredging them up from the past.

By contrast, *secondary memory*, in James's scheme, is "the knowledge of an event, or fact, of which meantime we have not been thinking, with the additional consciousness that we have thought or experienced it before." This type of memory "is brought back, recalled, fished up, so to speak, from a reservoir in which, with countless other objects, it lay buried and lost from view." With secondary memory, the information is no longer milling around the hotel lobby, but is instead resting upstairs and must be found and retrieved.

Remarkably, James's categorization of memory appears to have come solely from his own introspection. He did not conduct memory experiments on himself or others, although he may have talked with colleagues who did. After he proposed this scheme, however, scientists went to their laboratories to devise behavioral experiments to tease apart these two memory processes. Their work resulted in the concepts of what is now called short-term memory—James's primary memory—and long-term memory—James's secondary memory.

If short- and long-term memory represent two distinct kinds of cognitive processing, then their biological foundations should also differ. In addressing this issue, scientists have asked two basic questions: do separate neural circuits support short- and long-term memory, and can we identify in the respective brain circuits structural changes that contribute to memory storage? Researchers have pursued these fundamental questions broadly, with insights from theoretical, cellular, and molecular levels of analysis.

One early advance in examining the dual-process theory of memory came from Penfield's colleague, neuroscientist Donald Hebb. For some time, scientists had known that functions of the brain—remembering, thinking, or controlling body movements—depended on communication among brain cells, *neurons*. A major function of neurons is to send electrical and chemical messages across a *synapse*, a miniscule space between two neurons, to other neurons waiting to receive the message. Understanding how to link a complex process such as memory to some measurable activity in neurons was difficult—and still is.

In 1949, Hebb speculated that the central difference between the two types of memory is that long-term memory is accompanied by a physical change in the connections between neurons, whereas short-term memory is not. He proposed that short-term memory is made possible when neurons in a particular circuit talk continuously to one another in a closed loop, like a conversation kept alive by a group of people standing in a circle. Long-term memory, in contrast, comes from lasting new growth on the neuron at the synapse. If short-term memory is like an oral conversation, long-term memory is like a written transcript of past communication that can be taken out and reread at will.[1]

In developing his theory, Hebb was likely inspired by the famous Spanish anatomist Santiago Ramón y Cajal, who spent his career observing nerve cells through a microscope. In the late 1800s, Ramón y Cajal had proposed that learning was linked to a physical outgrowth of a nerve cell at the synapse. Hebb likewise believed that the structural connections between two neurons change physically and grow stronger as learning progresses. The ability of synapses to change structurally offers a way to record information permanently for later use.[2]

Hebb's model was enormously influential. It bridged the wide gap between psychology and biology, linking the seemingly elusive process of memory with a tangible change in the brain. It also gave scientists a way to frame further experiments, and set the stage for major breakthroughs in memory research. Hebb's postulate lives on in academia today: every student of neuroscience can recite Hebb's rule, "Cells that fire together wire together."[3]

Years later, inspired in part by Henry's story, neurobiologist Eric R. Kandel took up the subject of the cellular neurobiology of short- and long-term memory. In the late 1960s, Kandel and his colleagues began to study an invertebrate with a simple nervous system, *Aplysia* (sea snail), to see how it transformed short-term memories into long-term memories. The researchers focused on two simple forms of implicit learning: *habituation*, the process whereby organisms cease to respond to conspicuous but unimportant stimuli after many exposures to them, and *sensitization*, the process whereby experiencing a powerful stimulus leads to an increased reaction to a later stimulus that otherwise would have elicited a weaker response. In our daily lives, these unconscious mechanisms operate in the background to protect us and keep us focused. In habituation, we learn to tune out music blaring in the apartment next door, and in sensitization, being bitten by a neighbor's dog makes us fearful and tense the next time we hear a dog bark.

To study these simple forms of learning, Kandel and his collaborators focused on the snail's gill-withdrawal reflex, which protects its breathing apparatus. The gill is usually relaxed, but when something touches its siphon, the tube that ejects liquid from the snail's body, the siphon and gill withdraw into an opening. Kandel and his colleagues trained this simple response to demonstrate habituation and sensitization. In one experiment, they repeatedly delivered a mild touch to the siphon; the snail eventually became habituated to this touch, and its gill-withdrawal reflex weakened. In another experiment, the researchers gave the same mild touch to the siphon, but this time they simultaneously shocked the snail's tail. In this case, the snail became sensitized and produced a strong gill-withdrawal reflex, even when the weak touch was unaccompanied by the shock. Habituation and sensitization lasted anywhere from one day to several weeks, depending on the training protocol.

Because of the simplicity of *Aplysia*'s central nervous system, Kandel and colleagues could anatomically map the neural circuitry of the gill-withdrawal reflex and identify the synaptic connections among the cells in this circuit. They then inserted electrodes and recorded the activity of individual sensory and motor neurons. These experiments were possible because in *Aplysia* the cells are relatively large—up to one millimeter in diameter at the cell body. The electrophysiological recordings made it possible to identify the sensory neurons that were activated by touching the siphon and the motor neurons that initiated the reflex. With these methods, Kandel showed that learning was related to an increase in the electrical strength of the connections at synapses, with the result that one cell could communicate more effectively with its targets. This important study was one of the first to draw attention to the signaling properties of neurons, the cellular and molecular biology of how the connections between neurons are affected by learning.[4]

Crucially, in the same series of experiments, Kandel and his colleagues demonstrated that the mechanisms underlying short-term memory and long-term memory are different. Short-term memory, they learned, is associated with changes in synaptic function, as opposed to structure. During the course of learning, existing connections may become stronger or weaker, without any ostensible change in structure. By contrast, long-term memory requires physical changes at the synapse. Long-term memory requires protein synthesis; short-term memory does not. Kandel's experiments supported and amplified Hebb's insight that two distinct memory processes exist side by side.

Hebb had put the concept of *synaptic plasticity* on the table by proposing that the repeated stimulation of a group of neurons during learning gradually strengthens their connections, and in doing so establishes lasting memories. Twenty years later, Kandel provided strong support for Hebb's theory when he linked the recurrent activity of single neurons to particular kinds of learning in *Aplysia*. His discovery that different proteins are necessary for short- and long-term memory represented an important initial step toward uncovering the molecular basis of memory. Following in Hebb's and Kandel's footsteps, neuroscience researchers today focus on identifying the proteins and genes that will tell us how cells talk to one another and support learning.[5]

Discovering the molecular machinery that supports short- and long-term memory was crucial to understanding the roots of Henry's amnesia.

Similarly, examining his amnesia in behavioral studies became an opportunity for researchers to investigate how these two different types of memory are organized in the human brain. If the single-process theory were correct, then Henry's short-term memory should have been compromised. As it was, his short-term memory remained intact, while his long-term memory disappeared, suggesting that not only were they separate processes, but they also depended on different areas of the brain.

Having studied under Hebb, Milner was influenced by his dual-process theory of memory. She saw that Henry could provide experimental evidence to address the single-process versus dual-process debate. During Henry's 1962 visit to Milner's lab, her graduate student Lilli Prisko gathered data on his short-term memory. She asked Henry to compare two simple, nonverbal stimuli that were separated by a brief delay; the challenge for him was to hold the first item in his memory long enough to say whether it was the "Same" or "Different" from the second item. Prisko chose several different kinds of test items, enabling her to collect data from Henry in multiple experiments. Drawing conclusions based on a single experiment or task is risky, so Prisko avoided that problem by assessing Henry's short-term memory with complementary tasks. Some used sounds such as clicks and tones, and others used visual images such as light flashes, colors, and non-geometric nonsense figures. She also intentionally chose stimuli that were difficult to verbalize. In testing memory for colors, for instance, Prisko could not have used patches of red, orange, yellow, green, blue, and violet as test stimuli because Henry could have repeated the color names to himself during the delay intervals and outsmarted the test. Instead, she chose five different shades of red to minimize the possibility of verbal rehearsal.[6]

Henry lay on a stretcher in a quiet, dark area, separated by a screen from where Prisko sat in the main laboratory. They were the only ones present. She called out "First one" to indicate that a trial was starting. The stimulus on this trial was a series of flashes from a strobe light at a rate of three per second. After a brief delay, another set of flashes appeared, moving faster now at about eight per second. Henry had to say "Different" to indicate that there had been a change between the two stimuli. On other trials, the flashes appeared at the same rate, and he had to say "Same."

Testing Henry in this experiment was challenging, particularly at the beginning. Sometimes he talked instead of sitting quietly, needed prompting to give his responses, or responded after the first stimulus rather than waiting for the second. Every few minutes, Prisko repeated the instructions so that Henry would know what he was supposed to do. She also had to repeat several botched trials to complete the experiment.

The results of the first experiment provided an important basis for understanding Henry's capacity for perceiving and retaining information. He could easily and accurately perform the task when there was no delay between stimuli, making only one error in twelve trials. He clearly had no problem with the instructions or perceiving the test stimuli: he was perfectly capable of appreciating the difference between them when they were close together in time. With that knowledge, Prisko could assume that any problems Henry encountered during trials with longer delay intervals were a direct result of an inability to remember.

Prisko next tested Henry with the same sets of flashes, this time separated by fifteen, thirty, or sixty seconds. The ability to differentiate between stimuli becomes harder for everyone as the gap between them gets longer and as short-term memory weakens. In Henry's case, however, the differential was extreme. With a fifteen-second interval, Henry still performed well, making two errors out of twelve trials; when the delay was thirty seconds, his errors increased to four. At sixty seconds, his answers were wrong on six of the twelve trials, which is considered chance performance, no better than if Henry had flipped a coin each time to choose his answer. In comparison, normal control subjects made an average of one error out of twelve during the sixty-second delay between stimuli, even if Prisko distracted them.

The abrupt breakdown in Henry's performance showed that his short-term memory lasted less than sixty seconds. Somewhere between thirty and sixty seconds, his memory of what he had seen or heard crept away. With shorter delays, he performed better than chance: his mind held onto test stimuli, as long as they occurred within this narrow sliver of time. Henry's results were consistent with those of F.C. and P.B., both of whom Prisko tested soon after. They had fewer errors than Henry, but showed the same pattern of increasing errors as the time between the presentation of the two stimuli was extended, causing the memory of the first to fade away.

To Prisko's surprise, Henry showed some ability to retain certain kinds of information. After the test with flashes, she let him rest for a few minutes and then began the next task, this time using audible clicks. By now, Henry seemed to have improved as a test subject; although he still talked, he no longer called out after the first stimulus. She gave him a one-hour break before the next test, which was with colors. When he returned, Henry had completely forgotten who she was. Yet after being given his instructions, he seemed to understand the testing scheme better, talked less, and followed the instructions correctly. When she tested him again the next day, it was enough to give Henry the instructions once at the beginning of each new test. His scores were just as bad, and he had no memory of having done the test before, but somehow Henry knew what was expected of him.

How was it that Henry was able to learn the correct procedures—the "how to do it"—but remained unable to retain for more than a few seconds what the specific test stimuli had been? In 1962, no one could explain this strange distinction, but it gave all of us in Milner's lab a sense that Henry had a great deal to teach us about the nature of memory and learning.

Prisko's results from testing dealt a blow to the theory that memory was a single process. Around the same time, another case was emerging that posed a similar challenge. A patient in England, known by his initials K.F., sustained massive damage to the left side of his head and brain in a motorcycle accident. He was unconscious for ten weeks, and although he showed gradual improvement over the next few years, he began to have seizures. Like Henry, K.F. had a massive memory deficit, but of exactly the opposite pattern; remarkably, he could form new long-term memories despite his lack of short-term retention. He had a digit span of just two, and could repeat only one number, letter, or word reliably. If an examiner spoke pairs of words at the rate of one word per second, he could repeat both words correctly only half of the time. His short-term store had a very limited capacity. Nevertheless, K.F. performed normally on four different long-term memory tests, indicating that his long-term store for this information was intact.[7]

Taken together, the findings for Henry and K.F. indicate the existence of two independent memory circuits serving short-term memory and long-term memory, respectively, strongly supporting the dual-process theory. The two circuits have different anatomical locations: cortical processes mediate short-term memory, and medial temporal-lobe processes modulate long-term memory.[8]

K.F.'s case suggested that short-term memory processes are based in the cerebral cortex, the outer layers of the brain. Scoville did not touch this part of Henry's brain, so all of Henry's cortical functions were preserved, available for holding information online for brief periods of time, thus sparing his short-term memory. Further research has shown that short-term memories are scattered in different parts of the cortex, depending on the type of information they represent. Increasing evidence indicates that different areas of the brain are responsible for temporarily retaining memories of faces, bodies, places, words, and so on. These memories are not distributed randomly; instead, they tend to cluster near areas related to how the information was first perceived. The right parietal lobe, for example, is devoted to spatial abilities, so short-term memories related to spatial knowledge are maintained in that area. Similarly, the left side of the brain controls language, and short-term verbal memories are rooted predominantly in the left side of the cortex. With a better understanding of short-term memory as a separate process, we can explore in depth what happens in this brief period of time. We now know much more about the content, capabilities, and limits of short-term memory.[9]

Information lingers in short-term stores for less than a minute, but we can maintain information indefinitely by rehearsing it in our thoughts. Rehearsal effectively refreshes short-term traces, making them new again.

This is a good example of a *control process*. Control processes underlie our ability to manage thoughts in the interest of achieving a goal. We engage these processes constantly in everyday life, helping us focus on the task at hand, switch from one task to another, and block out unwanted intrusions.[10]

Imagine a businessman on a flight from Boston to San Francisco, connecting with a flight to Honolulu. After the plane lands in San Francisco, the flight attendant announces the gate numbers for flights to connecting cities. The man listens attentively as each city is called, and once he hears "Honolulu" and the gate number, he starts rehearsing it, repeating it over and over as he leaves the plane and successfully navigates through the crowds to the designated gate, willfully ignoring distractions along the way so that he does not forget where he is going. His gate number will likely disappear from his mind once he has boarded his flight; he kept it alive in his short-term memory just long enough for it to serve its purpose. We regularly juggle complex processes like these to help us remember and guide our behavior to achieve our goals.

Because Henry could rely only on short-term memory, he harnessed cognitive control processes to compensate for his memory deficit. By mentally rehearsing information he was asked to remember, he could sometimes keep thoughts fresh in his mind until asked to retrieve them. Milner noticed this ability during her first testing session with Henry in 1955 in Scoville's office. She gave Henry these instructions: "I want you to remember the numbers *five, eight, four.*" She then left the office and had a cup of coffee with Scoville's secretary. Twenty minutes later, she returned and asked Henry, "What were the numbers?"

"Five, eight, four," he replied. Milner was impressed; it seemed Henry's memory was better than she realized.

"Oh, that's very good!" she said. "How did you do that?"

"Well, five, eight, and four add up to seventeen," Henry answered. "Divide by two, you have nine and eight. Remember eight. Then five—you're left with five and four—five, eight, four. It's simple."

"Well, that's very good. And do you remember my name?"

"No, I'm sorry. My trouble is my memory."

"I'm Dr. Milner, and I come from Montreal."

"Oh, Montreal, Canada," Henry said. "I was in Canada once—I went to Toronto."

"Oh. Do you still remember the number?"

"Number? Was there a number?"

The complex calculations Henry had devised to keep the number in his head were gone. As soon as his attention was diverted to another topic, the content was lost. His forgetting that he had been rehearsing a number is unusual, but even people with intact brains lose information when they are distracted. Consider the airport example: if, while making his way through the airport, the businessman were distracted by a breaking news story on a television monitor, he would likely forget the gate number he had been working to retain in his short-term memory. If he did remember the gate number after being distracted, it would be because he drew on the resources of his long-term memory—a capacity that Henry lacked.[11]

Henry relied on his short-term memory in collaboration with his control processes. In conversation, he seemed normal because he could easily answer a question that was just asked of him. In this way, he could maintain a seemingly easy back-and-forth, staying sure-footed as long as nothing distracted his attention. He could keep names, words, or numbers in mind for a few seconds and could repeat this information, but only when the memory load was small and nothing else intervened to wipe the slate clean. If I were talking with Henry and another person started a different conversation with him, not only would he forget what I had just told him, he would also not remember that I had told him anything at all.

After I started my own lab at MIT in 1977, we had the opportunity to test the effect of distraction on Henry and four other patients with amnesia from other causes, all of whom had significant defects in their long-term memory and had only their short-term memory to rely on. We gave them the Brown-Peterson distractor task, which tests how quickly subjects forget information they have just absorbed. They look through a viewing window and see pairs of consonants followed by pairs of digits, with each pair visible for about three quarters of a second. The subjects might see, for example, *VG* and then *SZ*, followed by *83* and *27*. They read the consonants and digits, but are asked to remember only the consonants. Because the subjects are busy reading the digits, they cannot rehearse the consonants. By preventing rehearsal, the task measures how much information people forget and how quickly they forget it during a period of about five seconds, before being asked to recall the letters. When psychologist John Brown first introduced the distrator task in 1958, he found that healthy people could recall only one pair of consonants accurately. Subjects forgot the other information—the second pair—in less than five seconds when they could not rehearse it.[12]

In 1959, psychologists Margaret and Lloyd Peterson built on Brown's experiment by studying how accuracy changes when delay times are manipulated. In their version of Brown's test, the examiner said three consonants, such as *MXC*, and then a three-digit number, such as *973*. Participants had to count backward from the number by threes—*973, 970, 967, 964*—until given a signal to repeat the three consonants, *MXC*. They received the signal after different periods of time—three, six, nine, twelve, fifteen, and eighteen seconds. Peterson and Peterson found that the more time the participants spent on the distracting activity of counting backward, the fewer consonants they recalled. After the fifteen- and eighteen-second delays, they could remember almost nothing. This study showed that short-term memory lasts less than fifteen seconds under the influence of distraction.[13]

In the early 1980s, my lab modified the Brown-Peterson task to discover how long Henry's short-term memory lasted. This experiment was part of a broad investigation aimed at dissecting the component processes of *declarative memory*—a system that supports long-term memory for events and facts. When Henry and four other amnesic patients performed the Brown-Peterson task, they were neck and neck with healthy control participants without memory deficits at the three-, six-, and nine-second delays. At fifteen and thirty seconds, however, because their short-term store was full to capacity, the patients' scores fell well below those of the controls. In this experiment, normal people tapped into their long-term memory to retrieve information that needed to be held beyond about fifteen seconds, but people with amnesia could not.[14] This study helped define the limits of short-term memory.

Our understanding of short-term memory has become more complex as we study how we use our memories in everyday life. As we take in information from the world, we engage numerous complex processes in the brain. If someone tries to multiply sixty-eight and seventy-three in her head, she is performing calculations, saving the results, combining numbers, and checking their accuracy. The task requires much more effort than just regurgitating items held in her short-term memory; it is mental labor. She draws on the abstract ideas of numbers and multiplication, applying that knowledge to the problem at hand. This sort of process is called *working memory*: an effortful extension of short-term memory, a mental workspace where cognitive operations take place.

How does working memory differ from short-term, immediate memory? Think of short-term memory as simple and working memory as complex. Working memory is short-term memory on overdrive. Both are temporary, but short-term, immediate memory is the ability to reproduce a small number of items after a short or no delay (like saying 3–6–9), whereas working memory requires storing small amounts of information and simultaneously working with that information to perform complex tasks (like multiplying 3 x 6 x 9 in your head). When we engage short-term memory, we simply repeat a limited amount of information, whereas when we access working memory we can monitor and manipulate that information any way we want. Working memory organizes whatever cognitive and neural processes are necessary to achieve a short-term goal—deciphering long sentences, solving problems, following the plotline of a movie, carrying on a conversation, keeping track of a baseball game play by play.

Although the concept of working memory was first introduced in 1960, it was not until the 1980s that articles about working memory began to appear in the neuropsychological literature. In 1962, however, Milner administered a problem-solving test to Henry that we later realized measured not only his problem-solving ability but also his working memory capacity. Milner placed four cards on a table, side by side, and told Henry, "Here are your key cards." The first card had a red triangle, the second two green stars, the third three yellow crosses, and the fourth four blue circles. She told him to take a stack of 128 cards and place each card on the table in front of one of the key cards, wherever Henry thought it should go. After he placed a card down, she would say "Right" or "Wrong," and he was supposed to use this information to make as many correct choices as possible. He started sorting the cards, and initially she said "Right" if he matched a card to a key card based on the color and "Wrong" if he matched it by shape or number. After ten correct choices, she changed the sorting category without telling him. Now she said "Right" if he sorted by shape. She then changed again, using number as the sorting principle. Henry completed the task well and made few errors. He was engaging his working memory: he had to keep his attention focused on the correct category while he placed the cards on the table, listened to Milner's response, and decided on the next card placement accordingly. But despite his superior performance, he did not remember at the end of the test that he had been changing strategies— placing the cards first to match the color, next to match the shape, and then to match the number—in response to Milner's cues.[15]

The card-sorting task shed additional light on what Henry could and could not do. His ability to stick with a category as long as Milner said "Right" and to shift to another category when she said "Wrong" attests to his capacity to pay attention throughout a long test, to discriminate among different shapes and colors, and to think and respond flexibly. All these computations occurred online without the need to tap into long-term memory. When the test was over and Henry tried to think back over the entire test and recollect all he had just done, he was at a loss. The critical long-term memory traces were nowhere to be found.

Around the same time, Milner had the opportunity to administer the card-sorting task to a patient in whom Penfield had removed the front third of both frontal lobes to alleviate epilepsy. This man sorted all 128 cards based only on shape, even when Milner repeatedly said "Wrong"— an extreme example of perseverative behavior, responding in the same way over and over again. This case and many others that Milner tested with removals of either the left or right frontal lobe showed convincingly that the flexibility of planning and thought required for the sorting task depended on normal frontal-lobe function. Based on these findings, we can state with confidence that Henry's frontal-lobe capacities were excellent.[16]

In the 1990s, my lab dedicated a substantial effort in our studies with Henry to assessing his working memory. We expected that his working-memory capacities would be unaffected because he could monitor and manipulate items in his intact short-term stores. But the working memory tests sometimes posed a challenge for Henry in two respects. On one task, he had to respond quickly to keep up with the imposed pace of the test, and in some cases, he may not have had sufficient time to decide on and execute the correct response. On another test, the number of stimuli he had to monitor and manipulate exceeded the capacity of his immediate memory, and therefore, required engagement of long-term declarative memory, which he lacked.

For the test with time constraints, the N-back test, the stimuli were patches of color (e.g., *red, green, blue)*, presented one at a time on a computer screen at a rate of one color every two seconds. We asked Henry to press one button whenever the current color matched the color that had appeared immediately before, and a different button when it did not match. After he completed a number of such trials, the task became more challenging because we instructed him to press the button whenever he saw a color that matched the one that occurred two before it (i.e., with one intervening stimulus) and a different button when the color two before did not match.

On the N-back test with colors, Henry's performance was intact, attesting to the integrity of three key cognitive processes—*information maintenance* (Henry had to hold the colors online as they appeared), *information updating* (he had to constantly update the color he was holding online), and *response inhibition* (he had to inhibit the tendency to always press the non-target button because matches occurred less frequently than non-matches). When the stimuli were colors, the two-second time limit did not handicap him.

We later administered two similar N-back tests using six spatial locations and six meaningless shapes instead of colors. With these new stimuli, Henry's performance was no better than if he had been guessing. He had great difficulty maintaining the spatial locations and meaningless shapes in his working memory perhaps because he could not quickly attach verbal labels to them and respond correctly during the two-second window. Faced with test stimuli that were difficult to verbalize and the need to respond quickly, Henry was unable to succeed on this task.

On another test, self-ordered choosing, we measured Henry's ability to plan and keep track of a sequence of his own responses. He saw six designs displayed in a grid on a computer screen with three in the top row and three in the bottom row. We asked him to choose one design. Then he saw a new screen with the same six designs, each presented in a different position. This time he had to choose a different design. On the four subsequent trials, he again saw the six designs, each time in a different location, and again had to choose one he had not selected before. Henry completed this procedure three times in succession, and his performance was comparable to that of a control participant. In subsequent testing, however, when we increased the number of designs to eight and then twelve, Henry made more errors than the control. Because his errors tended to occur in the later trials, he could have been getting interference from constantly monitoring his experience of the early trials. Moreover, the demand of keeping track of eight and then twelve items, likely exhausted the limits of his immediate memory, and he did not have long-term declarative memory to fall back on.

Henry's limited performance on our working memory tasks does not lead to the conclusion that working memory depends on an intact hippocampus. He was thwarted by the tests that required him to respond rapidly or to engage his long-term memory. When the capacity of immediate memory is insufficient to hold the number or complexity of stimulus items online, declarative memory and the medial temporal-lobe circuits that support it mediate successful performance. When these circuits are dysfunctional as in Henry and other amnesic patients, they are likely to fail on challenging working memory tests. In 2012, memory researchers at the University of California, San Diego, reviewed ninety articles in the neuroscience literature on this topic and arrived at the same conclusion—when the requirements of a task are greater than the capacity of working memory can accommodate, performance is supported by long-term, declarative memory.[17]

Research on working memory has become a vast area of investigation, with more than twenty-seven thousand scientific papers on this topic. Ongoing studies in thousands of laboratories continue to dissect working memory processes and circuits to establish brain-behavior correlations in animals and humans. Because working memory relies on multiple cognitive processes—attention, impulse control, storing, monitoring, ordering, and information manipulation—it recruits multiple brain circuits in parallel. Consequently, working memory is highly vulnerable to neurological conditions, so we see working memory deficits in patients with attention-deficit/hyperactivity disorder, autism, Alzheimer disease, Parkinson disease, HIV, and strokes, and even normal, healthy aging. Individuals in these groups sometimes have trouble engaging working memory to achieve their goals because this challenging mental effort requires the brain to be fit; even subtle abnormalties can cause performance to suffer.

The notion of working memory emerged not from neuroscience labs but rather from the field of applied mathematics. Norbert Wiener, widely considered the most brilliant American-born mathematician of his generation, proposed in 1948 that the brain is like a computing machine. With this insight he established the discipline of *cybernetics*, the study of control processes in humans and machines.[18]

The metaphor of the human brain as an information processor had a far-reaching effect on the field of neuroscience. It influenced George A. Miller, Eugene Galanter, and Karl H. Pribram, three mathematically oriented thinkers at Stanford's Center for Advanced Study in the Behavioral Sciences, to unite the disciplines of cybernetics and psychology in their

1960 book, *Plans and the Structure of Behavior*. In this pioneering work, they argued that behavior must be directed by an overarching "Plan." They made a radical proposal that the brain could be compared to a computer, and the mind to a computer program, or Plan.[19]

Plans and the Structure of Behavior introduced the term *working memory*, a concept that quickly became an active area of research in cognitive science and cognitive neuroscience. This vast area of research goes beyond memory formation by addressing our ability to achieve complex goals. Specific goal-directed plans, analogous to computer programs, are stored somewhere and retrieved while they are being executed. "The special place may be on a sheet of paper," they wrote. "Or (who knows?) it may be somewhere in the frontal lobes of the brain. Without committing ourselves to any specific machinery, therefore, we should like to speak of the memory we use for the execution of our Plans as a kind of quick-access, 'working memory.' As Milner's card-sorting study with Henry demonstrated, not only were the authors spot-on in their characterization of working memory, they were also correct in their guess that it resides in the frontal lobes of the brain. We now know that the prefrontal cortex is critical for holding multiple thoughts in mind to construct and carry out a plan, just as Henry did when he performed the card-sorting task.

Over the next decade, the study of working memory exploded, as psychologists and neuroscientists sought to dissect the underlying cognitive and neural processes. In 1968, psychologists Richard Atkinson and Richard Shiffrin described a detailed model for human memory in "Human Memory: A Proposed System and Its Control Processes," still one of the most-cited writings in the literature on human memory. They divided memory into three stages: the sensory register, the short-term store, and the long-term store. The sensory register is the first entry point for incoming information from the senses. The information resides there for less than a second and then decays. In their single-process model, the short-term store is working memory; it receives input from the sensory register as well as from long-term stored memories. Information flows along a continuum, from the short-term store to the long-term store, which is a relatively permanent silo.[20]

Although the Atkinson-Shiffrin single-process model was influential, it did not fully account for the mechanism of long-term memory formation. If the model were correct, then Henry should not have had amnesia because, with the passage of time, information present at the short-term stage would have flown automatically into the long-term stage. Clearly, this did not happen—Henry's brain could not translate information from short-term processing mechanisms to long-term processing mechanisms. Still, Atkinson and Shiffrin's model is noteworthy for defining the short-term store as the individual's working memory where *control processes* come into play. Control processes vary from individual to individual. We decide what to pay attention to; rehearse information to keep it in the short-term store; and create mnemonics, such as *Every Good Boy Deserves Fudge*, which music students invoke to recall the notes on the lines of the treble clef—*EGBDF*. The Atkinson-Shiffrin model started the quest to understand the strategies that influence the processing of information held in working memory.

In 1974, psychologist Alan Baddeley and his colleague Graham J. Hitch posited that working memory is not a unitary system. They proposed that it consisted of three subsystems: a *central executive* that calls the shots, and two *slave systems* that do the hard work—one devoted to visual information and one to language. This model generated an explosion of experiments that tried to identify the mechanisms operating within each subsystem, how the transient processes interact with long-term memory, and the brain areas that are recruited during the performance of working memory tasks. Scientists have studied working memory in healthy people of all ages, twins, bilinguals, menopausal women, the congenitally blind, smokers, insomniacs, people under stress, and people with numerous neurological and psychiatric disorders. These experiments have had wide-ranging impact, with implications for education, evaluating treatments, training programs and their utility, and evaluation of psychiatric disorders.[21]

In recent years, however, Baddeley's model, which relies on dedicated holding areas—one for vision, one for hearing, and one for unique events—has been superseded by the concept of a more dynamic system. Current thinking considers working memory processes as interacting with long-term memory stores. By this view, information in working memory is kept alive by active processing in multiple areas within the temporal, parietal, and occipital lobes, in the same specialized brain areas where the initial perception of that information occurred. So, the circuits that are called into play when we first hear a name, see a face, or enjoy a landscape are the same ones that are active when we later remember that name, face, or landscape. The network that is recruited by working memory processes at a particular time depends on the content of our working memory and what we are trying to accomplish.

Twenty-first-century models of working memory emphasize the interactions between short- and long-term memory. Cognitive neuroscientists Bradley Postle, Mark D'Esposito, and John Jonides have each championed the view that working memory integrates specific information from different time periods—the sights, sounds, smells, tastes, and skin and body sensations that just entered the brain, and the contents of long-term memory that are relevant to these inputs. For example, if we try to multiply thirty-six by thirty-six in our heads, a task that engages working memory, we need to access our stored knowledge about numbers and multiplication to do the calculation. These researchers see working memory as an *emergent phenomenon* that arises from cooperation among many brain areas. As a result, the human brain is capable of multitasking and dealing with different kinds of information concurrently, switching from one task to another with great flexibility.[22]

Imagine a woman in a restaurant who is listening to a waiter list the day's specials. She keeps the list of dishes active in her working memory while simultaneously evaluating each dish, based on knowledge stored in her long-term memory. After shuffling these options in her head, she decides *not swordfish, because of its mercury content; not fried chicken, because it has too much fat; but the vegetarian pasta sounds similar to a dish I liked before*. She orders the pasta. Although she made this decision quickly, it came about through cooperation across networks of brain areas, enabling her to monitor and manipulate different kinds of information.

How does the brain accomplish this amazingly complex feat? In 2001, neuroscientists Earl K. Miller and Jonathan Cohen posited that the prefrontal cortex orchestrates thought and action to achieve internal goals, such as deciding what to order for dinner. The neural circuitry in the prefrontal cortex, which supports working memory, allowed the woman in our example to maintain the words she just heard, to experience visual images and tastes of the food, to retrieve memories of her recent meals, and to examine her knowledge and opinions about food. In short, her choice was guided by *top-down processing*, using her experience to guide decision-making.[23]

Top-down processes enable us to modulate the impact of diverse sensory information. These computations include planning a series of actions, managing goals, coordinating and monitoring automatic processes (such as reacting to a mouse in the kitchen), inhibiting powerful habitual responses, focusing attention selectively, and suppressing irrelevant sensory inputs so that we can generate internal representations of goals and how to attain them. The prefrontal cortex in the front of the brain guides the flow of information along pathways in the back of the brain and in areas under the cortex, which are crucial for problem solving and decision-making. One requirement of the prefrontal cortex is that it must be flexible—able to adapt to changes in the environment inside or outside of the body, and able to entertain new goals and procedures.[24]

Henry's working memory was sufficiently robust to allow him to play bingo, speak in sentences, and do simple arithmetic in his head. He was unable, however, to integrate his online thoughts with memories of the recent past. If he ordered a meal in a restaurant, he could make a choice based on what he liked and disliked before his operation, but he could not take into account what he had eaten the day before, whether he had to select low-calorie items to control his weight, or if he needed to limit his salt intake. Henry depended on his caregivers to fill in that information and much more. His daily life had many limitations because he lacked vital long-term memory capacities.

What is it like to experience life with only short-term memory to rely on? No one would doubt that Henry's experience was a tragedy, but he rarely seemed to suffer and was not continuously lost and frightened—quite the contrary. He always lived in the moment, fully accepting the events of daily life. From the time of his operation, every new person he met was forever a stranger, yet he approached each one with openness and trust. He remained as good-natured and pleasant as the polite, quiet person his high-school classmates knew. Henry answered our queries patiently, rarely getting angry or asking why he was being questioned. He understood his situation enough to know that he had to rely on others and willingly accepted help. In 1966, when Henry was forty, he visited the MIT Clinical Research Center for the first time. When asked who packed his bags for him, he answered simply, "It must have been my mother. She always does those things."

Henry was free from the moorings that keep us anchored in time, attachments that can sometimes be burdensome. Our long-term memory is critical to our survival, but it also hinders us; it prevents us from escaping embarrassing moments we have lived though, the pain we feel when thinking about lost loved ones, and our failures, traumas, and problems. The trail of memory can feel like a heavy chain, keeping us locked into the identities we have created for ourselves.

We can be so wrapped up in memories that we fail to live in the here and now. Buddhism and other philosophies teach us that much of our suffering comes from our own thinking, particularly when we dwell in the past and in the future. We replay moments and events that happened before and spin narratives about what might transpire in the future, becoming mired in the emotions and anxiety of these stories. Often our thoughts and feelings have nothing to do with the concrete reality of the present. When people meditate, they may attend closely to their breathing or to a particular body part, or they may repeat a mantra—whatever helps them maintain contact with the present moment and avoid getting caught up in distracting thoughts and narratives. Meditation is a method for training the mind to have a new relationship with time, knowing only the present tense, unburdened by the power of memory. Dedicated meditators spend years practicing being attentive to the present—something Henry could not help but do.

When we consider how much of the anxiety and pain of daily life stems from attending to our long-term memories and worrying about and planning for the future, we can appreciate why Henry lived much of his life with relatively little stress. He was unencumbered by recollections from the past and speculations about the future. As frightening as it seems to live without long-term memory, a part of us all can understand how liberating it might be to always experience life as it is right now, in the simplicity of a world bounded by thirty seconds.

Five

Memories Are Made of This

Our research with Henry focused on two kinds of investigation. One used brain-imaging tools to reveal the anatomy of his surgery and thereby show exactly what areas had been removed and what had been left behind. This level of detail is critical when neuroscientists are trying to relate the function in discrete brain areas to specific behaviors. The other kind of study harnessed cognitive tests to evaluate Henry's memory and other intellectual functions. We knew from Milner's testing in 1955 that his IQ was above average. Still, we wondered about other aspects of complex thought. In addition, it was important to assess his perceptual capacities to be sure that he was receiving accurate information about the world.

The roots of memory formation are planted in our sensory organs as collections of separate threads. If we focus on our environment at this very moment, we realize that we are receiving different kinds of information simultaneously through our eyes, ears, nose, mouth, and skin. We are perceiving sights, sounds, smells, tastes, and touches. These diverse pieces of information are automatically channeled along separate pathways to our cortex where they are processed in areas specialized for each sensory modality. This material also reaches our hippocampus where the various sensations come together and memory formation begins. The process of laying down a memory requires back-and-forth communication between our hippocampus and the areas distributed throughout our cortex where the sensory information was first perceived. During this interaction, the hippocampus organizes the cortical components of a memory so they are available for retrieval as a whole memory and not a bunch of disconnected fragments. Together these traces contain a rich representation of our experiences.

Because memory formation depends heavily on receiving valid information from the senses, it was necessary to establish the integrity of Henry's sensory functions. If he could not perceive photos of faces normally, how could we expect him to remember them? The same was true for the other senses. For this reason, we considered it important to evaluate Henry's sensory capacities and did so on and off from the 1960s through the 1980s. The evidence convinced us that his bad memory could not be dismissed as a side effect of impaired visual, auditory, or tactual perception.

The demonstration that Henry's memory could come to a standstill while his intelligence remained intact indicates that the capacity to form new memories is separate from overall intelligence. To establish conclusively that his acumen had survived, we conducted numerous experiments that examined Henry's higher intellectual functions, such as problem solving, orientation in space, and reasoning. A tremendous boon to this effort was that Henry enjoyed test taking and was a very attentive and cooperative research participant. His pattern of cognitive strengths and weaknesses, demonstrated over years of research, helped define the scope of the amnesic syndrome, and of his preserved capacities. Henry harnessed many different abilities to compensate, as best he could, for his tragic memory impairment.

When my colleagues and I first began studying Henry, we did not know exactly how much damage his brain had suffered as a result of Scoville's surgery. The only information we had came from Scoville's account of what he had taken out, which itself was only an educated guess. Over the next five decades, new technologies gradually emerged that enabled us to examine Henry's brain lesion in greater and greater detail.[1]

We knew from Henry's 1946 pneumoencephalogram that his brain appeared normal prior to his surgery. It took nearly half a century, however, to obtain an accurate picture of Henry's postoperative brain. Brain imaging took a major leap forward in the 1970s with the invention of computed tomography (CT), which uses X-rays and a powerful computer to create cross-section images of the brain. CT enables doctors and researchers to examine brain structures slice by slice, focusing images to one plane at a time and eliminating interference from surrounding structures.

In August 1977, I asked a colleague in the Department of Neurosurgery at Mass General to order a CT scan of Henry's brain. The radiologist observed surgical clips in the region of the temporal lobe on both sides of Henry's brain, deliberately left to control bleeding. Both temporal lobes were slightly atrophied (shrunken), and the Sylvian cisterns—the spaces between the temporal and frontal lobes that contain cerebrospinal fluid—were slightly enlarged on both sides, another indication of atrophy. Henry's cerebellum showed similar evidence of shrinkage. The images showed no sign of a brain tumor or other abnormality. The CT scan confirmed only that he was missing tissue deep in each temporal lobe, but we were unable to judge exactly which structures had been removed and to what extent.

By the mid 1970s, mounting evidence from animal and human tests had convinced scientists that the hippocampus was vital for converting short-term memory into long-term memory, but we still needed direct proof that the hippocampus was responsible for Henry's amnesia. Another CT scan, conducted in 1984, simply verified the results of the 1977 study. Because these scans showed just the spaces in his brain and not the anatomy of what remained, we sorely needed a better tool.

In the early 1990s, we were finally able to thoroughly assess the damage that had been done to Henry's brain, thanks to the development of magnetic resonance imaging (MRI), which had been invented in 1970. Commercial scanners became available in the early 1980s, and MRI evolved into a mainstream tool by the end of the decade. MRI is superior to CT in distinguishing one brain area from its neighbors. Like CT, MRI takes cross-section images, but instead of relying on radiation, uses radio waves and a powerful magnet to obtain precise images of tissues. The magnetic field forces hydrogen atoms to align in a particular way, while the radio waves bounce off the hydrogen protons in the body and produce a signal. Different types of tissues give off different signals, which a computer can re-create as black-and-white images.[2]

Using MRI scans, we could look through Henry's scalp and skull to see his brain. With this new method we could identify small brain structures and get a clearer picture of the damage than with CT scans. Before MRI, the only way to see the brain's anatomy in detail was to look at it directly, during either a surgical procedure or an autopsy. Henry's first MRI scan in 1992, at Brigham and Women's Hospital in Boston, was an exciting moment for all of us who had spent decades studying him. For the first time, we had a clear view inside what was perhaps the most studied brain in the world.[3]

Henry's brain generally looked normal for a man of sixty-six, with the exception of his cerebellum—the grooved bulb near the brain stem that supports motor control. In the 1960s, we could only infer this damage from abnormalities in his neurological examinations, but the MRI images showed a shriveled cerebellum, surrounded by extra space filled with cerebrospinal fluid. Although we knew that Henry's cerebellum was abnormal, we were struck by the extent of the atrophy. This damage was due to the drug-related death of neurons. For many years, Henry had taken Dilantin to prevent seizures, until the drug caused ringing in his ears. In 1984, his doctors replaced the Dilantin with a different seizure medication, but the tinnitus did not subside. Dilantin left him with other permanent disabilities—loss of feeling in his hands and feet, and challenges with balance and movement. Henry strolled from place to place in a slow, unsteady gait, his feet far apart for stability—symptoms of the cerebellar atrophy that was so striking in his MRI scans.

As the scans moved into the interior of the temporal lobe, we could see that the operation forty years before had left an irretrievable absence— two nearly symmetrical gaps in the middle of Henry's brain. Missing was the front half of each hippocampus and the areas that interface with the hippocampus—the entorhinal, perirhinal, and parahippocampal cortices. Also removed was most of the amygdala, an almond-shaped group of structures that supports emotion. The entire lesion in Henry's brain extended just over five centimeters from front to back, far less than the eight centimeters Scoville had estimated. Approximately two centimeters of hippocampus still remained in each half of the brain, but this residual tissue was useless: the pathways carrying information to it had been destroyed (see Fig. 3).

During the MRI scanning sessions, Henry was, as usual, an agreeable subject. He was not claustrophobic inside the scanner, and afterward we would serve him lunch—a sandwich, tea, and his favorite pudding or pie. Henry was a gourmand, and as he aged, his belly grew so rotund we worried about fitting him into the MRI's tubular scanner. After Henry came out of the scanner, my colleagues at the imaging center were always eager to have a conversation with him, so he often attracted a small group of fans. But he never questioned why he was such an attraction, taking it all in stride.

In 1993 and from 2002–2004, we performed several more MRI studies on Henry. By then, MRI analysis techniques had improved, and we could measure more precisely the amount of brain tissue that had been removed or spared. Once we had clearly defined the anatomy of Henry's brain lesions, we had the exciting opportunity to link his impairments to the damaged areas, and his good performance to the spared areas. MRI evidence strongly supported the conclusion that the medial temporal-lobe structures removed from Henry's brain are crucial for long-term *declarative* memory, the conscious retrieval of facts and events. Henry's memory was severely impaired, regardless of the kind of test (free recall, cued recall, yes/no recognition, multiple-choice recognition, learning to a criterion), stimulus material (words, digits, paragraphs, pseudo-words, faces, shapes, clicks, tones, tunes, sounds, mazes, public events, personal events), and the sensory modality through which information was presented to him (vision, audition, somatosensory system, olfaction). His impairment was not just severe; it was pervasive. This *anterograde amnesia*, which characterized his postoperative life, boiled down to deficient acquisition of *episodic knowledge*—memory for events that happened at a specific time and place—and of *semantic knowledge*—general knowledge about the world, including new word meanings.

An important finding in the MRI scans was that, beyond the hippocampus, a bit of medial temporal-lobe tissue was left behind on both sides, the back part of his parahippocampal gyrus (perirhinal and parahippocampal cortices). We entertained the possibility that this spared cortex, known from studies in monkeys to be important for memory, was engaged when, from time to time, Henry would surprise us by consciously remembering something he had no business remembering. He could draw the floor plan of a house that he moved into after his operation; he could recognize complex color pictures up to six months after he had studied them; and he could describe a few details about celebrities who had become famous after his operation. The perirhinal and parahippocampal cortices receive information from other cortical areas, and this stored information was likely used in building these memories. Evidence from experiments in animals and humans suggests that the different medial temporal-lobe structures act independently and can mediate behavior flexibly, via conversations with specific cortical-processing streams. These cortical mechanisms enabled Henry, in everyday life, to occasionally retrieve bits of stored information about the world.

The MRI images also revealed that the vast expanse of cortex on both sides of Henry's brain was normal. Thus, his cortical functions—short-term memory, language capacities, perceptual abilities, and reasoning—were undisturbed. In addition, circuits within Henry's healthy cortex and areas underneath supported several kinds of *nondeclarative memory*, skills and habits learned without conscious awareness. Henry's case showed us that these capacities are independent of the hippocampus.

I arrived at MIT in 1964 after completing my PhD at McGill University. I was a research scientist in what was then the Psychology Department. It was a growing department energized by scientists who represented disciplines ranging from neuroanatomy to psycholinguistics. The atmosphere was stimulating and collegial. Our chair, Hans-Lukas Teuber, was a German immigrant and an influential figure in the study of the brain. My own mission upon arriving at MIT was to establish a lab that focused on patients with neurological disorders. Over the years, the patient groups included veterans from World War II and the Korean War who had sustained head injuries, and patients who had undergone psychosurgery at Mass General. In investigating these groups of patients, I conducted a broad examination of their cognitive and motor functions, and my expertise grew beyond my PhD research, which had focused on the sense of touch. I was always particularly interested in memory, and in the late 1970s began studying patients with Alzheimer disease and other neurodegenerative disorders. In the 1980s, my colleagues and I expanded our study of aging to include alterations in the brain and related behaviors in healthy women and men. All along, my lab continued to study Henry intensively, interspersing studies on him with those on other types of patients.

The MIT Clinical Research Center (CRC), the base of all this testing, was established in 1964 as part of a larger movement to create federally funded centers for academic research on human diseases. During the administrations of John F. Kennedy and Lyndon B. Johnson, the federal government's role in healthcare expanded, and biomedical research was one of the beneficiaries. The resulting clinical research centers, funded by the National Institutes of Health, were instrumental in applying scientific techniques to studying disorders in a clinical setting. Our CRC was a small, ten-bed unit, situated on a single floor of an unobtrusive brick and concrete building on the MIT campus. It was designed with overnight accommodations for patients, allowing easy access to our testing rooms on the same floor.

The CRC became a home away from home for Henry; the CRC staff and the many researchers who passed through our lab became his extended family. Henry was admitted to the CRC for testing fifty-five times from 1966 to 2000. He sometimes stayed for three weeks or a month when lab members administered a spectrum of learning tasks that required several consecutive days of training.

Henry traveled to and from the CRC by car, with myself or another MIT researcher driving. A second person always accompanied the driver in case Henry had a seizure or some unforeseen event occurred. On these two-hour drives, he passed the time looking out the window, and sometimes the pictures on billboards triggered monologues that recurred from trip to trip.

Howard Eichenbaum, a collaborator from Wellesley College, recalls one such trip in 1980 when he drove to Bickford, Henry's nursing home, to transport him back to the MIT CRC. On the way there, Eichenbaum pulled into McDonald's for lunch, and returned to his car with a cup of coffee. When he arrived at Bickford, he went inside, talked with a staff member, and escorted Henry outside to his car. With Henry comfortably seated in the back seat, they set out for Boston. After a few minutes, Henry noticed the coffee cup on the dashboard and said, "Hey, I knew a fellow named John McDonald when I was a boy!" He proceeded to relate some of his adventures with the friend; Eichenbaum asked a few questions and was impressed by these elaborate childhood memories. Eventually, the story ended and Henry turned to watch the scenery passing by. After a few more minutes, he looked up at the dashboard and remarked, "Hey, I knew a fellow named John McDonald when I was a boy!" and proceeded to reproduce virtually the identical story. Eichenbaum again asked probing questions in an effort to continue the interaction and to determine whether the facts of the story would be the same. Henry was unaware that he was recounting the tale almost verbatim. After a few minutes, the conversation ended, and he turned to view the scenery again. Just minutes later, Henry again looked up to the dashboard and exclaimed, "Hey, I knew a fellow named John McDonald when I was a boy!" Eichenbaum helped him reproduce the same conversation yet again, and then quickly disposed of the cup under his seat.[4]

The nurses and kitchen staff at the CRC all doted on Henry. He had a private room with an adjoining bathroom, and every morning the nurses would wake him and help him get ready for breakfast. Around nine, he would begin testing in one of the well-equipped testing rooms down the hall. We would typically conduct multiple studies in parallel, with different lab members giving different tests in alternate sessions. So as not to tire Henry out, we gave him frequent breaks, often stopping for cookies and a cup of tea in the afternoon. The CRC dietitian and her staff served him homemade meals, preparing his favorite foods, such as French toast and cake. His only dislike was liver. Over the years, Henry—always a large man—grew a substantial belly, and though the dietician reined in his caloric intake, she always allowed him dessert. After lunch and in the evenings, he went to the lounge where he socialized with the other research participants, worked on puzzles, and watched movies.

In this ideal research environment, my colleagues and I had the wonderful opportunity to investigate the strengths and deficiencies of Henry's intellect. An early focus was on his perceptual capacities. Perception and memory are linked in that the information we perceive through vision, hearing, touch, smell, and taste provides the raw ingredients for our memories. All sensory modes contribute to memory formation, so we wanted to rule out any basic perception problems as culprits for Henry's poor memory. During his first admission to the MIT CRC in 1966, part of our plan was to extend the examination of his vision and hearing with greater precision than had been done in previous clinical neurological examinations. Brenda Milner came down to MIT from Montreal, and with the help of my colleague, Peter Schiller, we administered a broad spectrum of tests over the course of Henry's seventeen-day stay.

To confirm that Henry could see all areas in his visual field—those straight ahead, up, down, and off to each side—we asked him to place his chin on a chin rest and stare at a point directly ahead inside a bowl-shaped instrument. His task was to push a button each time a tiny light flashed in different parts of the bowl, while keeping his eyes glued to fixation point. With this method, we found that Henry's field of view was normal in all directions.

In a test of visual perception, *masking*, Henry saw a large letter on a screen followed immediately by a mask that covered the letter and stopped processing of the letter in Henry's visual circuits. The key measure was how much exposure time he needed to name the letter. In a second task, *metacontrast*, Henry saw a solid black circle for ten milliseconds and then, for ten milliseconds, a larger black doughnut whose inner edge touched the outside of the circle. If the circle and doughnut were flashed at the same time, Henry saw the two separate stimuli combined into a single large black circle. But if they were flashed one after the other with a tenth of a second between them, the circle disappeared and Henry saw only the doughnut. When the interval between the circle and the doughnut was increased to a second, Henry perceived the circle and doughnut as separate objects. Here the critical measure was the time that elapsed between the circle and the doughnut for Henry to see them as two separate objects. On both measures of visual perception, masking and metacontrast, Henry's performance again resembled that of control participants.[5]

We next tested Henry's ability to perceive more complex stimuli, such as faces and objects. We showed him forty-four black and white patterns, each suggesting a face. He responded quickly and accurately when asked to judge the sex and approximate age of each person. On another task, he had no difficulty identifying sketchy drawings of twenty objects (see Fig. 4).[6]

To assess Henry's hearing, we seated him comfortably inside a sound attenuated booth and asked him to wear headphones though which we could present tones to one ear or the other. He held a device with a button and depressed it when the tone came on and released when it went away. We placed written instructions in front of Henry, so he always knew what to do. A very faint, inaudible tone would come on and slowly become louder. As soon as Henry heard the tone, he pushed the button. Then the tone would gradually become softer, and Henry pushed the button when he could no longer hear it. By repeating this procedure at several sound frequencies, we showed that Henry's hearing was normal from low to high frequencies.

Establishing the integrity of Henry's sense of touch was trickier because many years of taking Dilantin had left him with a peripheral neuropathy—sensory loss restricted to the parts of his body that are covered by gloves and socks. On formal testing, he showed decreased sensitivity in these areas, but he could still identify common objects by touch and could appreciate the shape of patterns perceived with his hands well enough to construct replicas when given the necessary blocks to do so.

The exception to Henry's preserved sensory capacities was his sense of smell. Around the world, people delight to the smell of freshly baked bread, but after his operation, Henry could not enjoy, and therefore record, this heavenly sensation. The hippocampus does not support the sense of smell (olfaction), but several structures adjacent to the hippocampus do. When we inhale the smell of fresh-out-of-the-oven bread, we activate neurons that transfer olfactory information from the nose to the brain's major receiving areas for this sensation. These areas include the front part of the parahippocampal gyrus, part of the amygdala, and the cortex around the amygdala. Scoville's operation report indicated that he removed these key olfactory areas from Henry's brain. The operation spared other primary olfactory areas in the frontal lobes, so in 1983, we conducted several experiments to determine whether these parts of the brain, still intact in Henry, could support any olfactory perception.[7]

To test Henry's sense of smell, we asked him to sniff bottles containing a common odor such as coconut, mint, or almond, and to select the name of the odor from five choices written on a card in front of him. Although this was not a memory test, the only choice he got correct was distilled water; when he sniffed that bottle, he responded, "Nothing." His performance showed that he could detect the presence of an odor normally—odor versus no odor—but that his brain was not giving him any information about the nature of the odors. He could not name odors correctly or tell them apart: he could not say whether two odors presented consecutively were the same or different, and was unable to match a sample odor to one of two choices. Interestingly, he could assign names to odors, but the ones he chose bore no obvious relation to the actual odor, and he did not use them consistently. So when he smelled a bottle containing the odor of cloves, he responded, "fresh woodwork" on one occasion and "dead fish washed ashore" the next time. I have no idea what inspired these responses.[8]

To rule out the possibility that his deficit could be reduced to a general problem with naming, we showed that he could name foods when he handled them in a bag, using his sense of touch, or saw them briefly, using vision. One incident in particular captured the essence of Henry's lack of olfactory sense: he correctly identified a lemon by sight, then sniffed it and said, "Funny, it doesn't smell like a lemon."[9]

But Henry's operation did not completely eliminate his sense of smell. In addition to being able to detect the presence of an odor compared to distilled water, he also performed normally on an intensity discrimination task. This task measured his ability to distinguish between different strengths of a particular odor. The examiner asked Henry to sniff one sample, then another, and to choose the stronger of two. He correctly chose the sample with the higher concentration of the odor; he just had no idea what that odor was.[10]

The results of this single concerted study of Henry's olfactory perception propelled the science forward. The revelation for neuroscientists was that the brain circuit that is responsible for odor detection—*this bottle contains an odor*—and odor intensity discrimination—*this odor is stronger*—is separate from the circuit that supports odor discrimination—*this smells like cloves*. Henry's capacity to detect even weak odors, to differentiate odor samples based on their strength, and to adapt to a strong odor indicated that the machinery that carried olfactory information from the nose to the cortex was at least partially intact. In addition, it is possible that a pathway to other olfactory cortices in his frontal lobes above his eyes was spared, helping to support the preserved behaviors. Nevertheless, these remaining inputs were insufficient to maintain odor discrimination, demonstrating that medial temporal-lobe structures play a critical role in odor matching and identification. Thanks to Henry, we now know that odor discrimination takes place in the front part of the parahippocampal gyrus, the amygdala, and the cortex around the amygdala. This ability to discriminate one odor from another and to recognize specific odors depended on these areas removed from Henry's brain, whereas the more elementary processes of detection, adaptation, and intensity discrimination relied on separate networks that were undisturbed.[11]

Amnesic patients do not typically lose their sense of smell, and indeed the deficit in Henry's case was not part of his amnesia. His loss was due to the removal of brain tissue during his operation. The continuing postmortem examination of his brain will tell us definitively about the integrity of remaining olfactory circuits, about which we could only speculate during his life. Specifically, it will help us understand the structure and organization of pathways that progressed from his nose to cortical olfactory areas in his frontal and temporal lobes.

Knowing that Henry's perceptual abilities were normal, save for smell, we could confidently attribute his inability to remember information received through vision, hearing, and touch to a memory impairment, and not a failure to sense the test materials in the same way as healthy participants.

Once we ruled out sensory loss as an explanation for Henry's deficient memory for information received through vision, hearing, and touch, we could start to catalog the deficits related to his brain operation. We were beginning to understand the extent to which memory depended on a few centimeters of tissue in the medial temporal lobe—those that Henry lacked. Today, the role of the hippocampus in memory is well established, and for decades, Henry played a key role in advancing this knowledge. At the time, however, he was our guide in exploring uncharted territory.

The tragic outcome of Henry's operation inspired neuroscientists to create animal models of amnesia. Initial attempts in the 1960s and early 1970s to create a memory impairment like Henry's in monkeys and rats were unsuccessful. Animals with lesions in both hippocampi had little if any difficulty on standard memory tests. Researchers began to make progress during the late 1970s when they devised new and more challenging ways to test memory, requiring the animals to recognize complex visual stimuli or learn mazes. After scientists began recording the activity of single cells in the hippocampus, a popular theory proposed in 1978 held that the hippocampus played a key role in spatial memory, and that this neural activity resulted in the establishment of cognitive maps, mental maps of one's environment.[12]

Aware of this emerging evidence, Milner and I decided to test Henry's maze learning ability in 1962 when I was a graduate student in her lab. We wanted to examine Henry's memory with tasks that did not rely heavily on verbal stimuli, such as words and stories, because previous studies with Henry had already covered that ground. Pursuing this new direction, Milner and I explored his spatial-learning ability using two maze-learning problems, one executed with the use of vision and the other with the use of touch. First, Milner trained Henry on the visual maze for three days, and then I trained him on the tactual maze for four days (see Fig. 5a).

The visual maze, placed on a table, was a thirteen-inch square wooden board with a ten-by-ten array of bolt heads set one inch apart. Milner designated a path from the start in the lower-left corner to the finish in the upper-right corner. Henry had to discover this path by trial and error. He held a metal stylus in his right hand and proceeded, one step at a time, from bolt head to bolt head. If he took a wrong step, he heard a loud click from an error counter and had to go back to the previous bolt head. Eventually he reached the finish, completing the first training trial. On the first day of training, Henry completed seventy-five trials and did the same on each of the next two days, for a total of two hundred twenty-five trials. At the end of each trial, Milner recorded the number of errors and the completion time (see Fig. 5b).[13]

The tactual maze was 12.75 inches by 10 inches and had paths cut into an aluminum sheet that rested inside a wooden frame. Henry sat on one side, where a black cloth curtain covered the frame to prevent him from seeing the maze. I sat on the opposite side, which was open so I could observe his hand, the stylus, and the maze as Henry advanced through it. I introduced him to the task by asking him to put both hands under the curtain to feel the perimeter of the maze, and oriented him to the maze by guiding his right hand, holding a stylus, to the start, next to the finish, and then back to the start. I then instructed him to move the stylus along the paths to find the correct route from start to finish. Each time Henry entered a blind alley, I rang a bell signaling that he should back up and try another path. Henry completed two sessions of ten trials each on four consecutive days, and I noted his errors and completion time, trial by trial.[14]

In these 1962 experiments at the Neuro, Henry failed on the visual maze and the tactual maze to reach the criterion of learning—three consecutive errorless runs. Even after completing many more trials than our control subjects needed to learn the correct route, he showed no improvement. Taken together, these experiments demonstrated that the deficit in maze learning was not restricted to a single sensory mode because it was evident when the task was done with visual guidance and also when visual guidance was completely excluded.

In 1953, when Henry returned home from the hospital after his radical operation, it became clear to his parents that even mundane activities would be a challenge for him. His boss at Royal Typewriter in Hartford must have liked Henry and been satisfied with his work before the operation because he allowed Henry to resume his assembly-line job afterward. But soon the boss phoned Mrs. Molaison and told her that Henry was too forgetful to do his job. He still had a sense of what his work entailed, but lacked the specific declarative knowledge to carry out his assignment—even though it was the same task performed over and over. Now unemployed, Henry stayed home with his parents, under his mother's constant care. Singlehandedly, she looked after all of his needs for the next three decades. Henry defined her life.

Henry helped his parents with household chores but forgot the locations of items he used frequently. His mother had to remind him where to find the lawnmower even if he had used it the day before. He could not do anything away from the house by himself, including going for a short walk. He would read the same magazines repeatedly, and would complete jigsaw puzzles without realizing he had already done them.

Ten months after the operation, Henry's family moved to a different house in East Hartford, just a few blocks away on the same street. The change was drastic for Henry. He could not learn his new address and would could not guide a driver to his house. His spatial memory—declarative memory for spatial locations—was deficient.

Four years later, in 1958, the family purchased an eight-hundred-and-sixty-square-foot bungalow at 63 Crescent Drive in East Hartford. By all expectations, Henry should have failed to remember this address too. Instead, he greatly surprised us. During a 1966 visit to MIT, Henry knew this address and was able to draw an accurate floor plan of the house from memory. Even more astonishing, in 1977, three years after he moved out of that house, he still responded with "63 Crescent Drive" when I asked where he lived, and again drew a floor plan, sketched in hesitant lines but with doors marked and rooms labeled. I contacted the person who then lived at 63 Crescent Drive and obtained the floor plan. The layout matched Henry's drawing, and he was able to recite this one address throughout the rest of his life.[15]

It was remarkable that Henry could remember the plan of a house he had never seen before his operation. Walking from room to room, day after day for sixteen years allowed him to construct a mental map of the house over time. But his knowledge was more than a vague sense of what was where. For example, he could picture his house in his mind's eye and tell me which way to turn to get from his bedroom to the bathroom, and where the front and back doors were located. His ability to recall the address in conjunction with the layout suggests that this home had become part of his world knowledge; this was information he should not have been able to learn.

Acquiring the floor plan of 63 Crescent Drive occurred without conscious awareness of the learning process, and with Henry's attention focused on other things. Habit learning is also unconscious, but habits are noncognitive—automatic, involuntary, and inflexible. Henry's spatial knowledge of his house was cognitive. He could use his spatial knowledge to picture the rooms in his house voluntarily, in relation to one another, and to describe consciously the route from point A to point B. This flexibility in navigating an internalized spatial map is markedly different from a habit.

Not until we saw the MRIs of Henry's brain did we understand his remarkable capacity to draw the floor plan. By the 1990s, scientists had uncovered a network of brain regions, including the hippocampus and areas in the cortex, which are engaged when remembering the topography of spaces. Once we could see the precise structures that had been removed or spared in Henry's brain, we discovered that some components of this brain network for processing information about space were still present. They included specific areas in the parietal, temporal, and occipital lobes—the somatosensory cortex, parietoinsular vestibular cortex, visual cortex, part of the posterior parietal cortex, inferotemporal cortex, and posterior cingulate/retrosplenial cortex.[16]

Clearly, enough tissue remained to form a memory of the house he had navigated countless times for years—a depth of exposure that we had not captured in our tests of Henry's ability to learn. In navigating the rooms of his house every day, he learned via the same process of immersion that someone trying to learn a foreign language might engage. Simply by following his daily routine, he enriched his mental map in small increments day after day after day—a perfect example of learning by mere exposure.

This astonishing evidence of spatial knowledge, acquired slowly over time, raised the intriguing question of whether this ability would extend to a test of spatial orientation in the laboratory. Henry's amnesia was not a detriment in this investigation because this spatial task did not rely on his long-term memory. We sought to discover whether, without a functioning hippocampus, Henry could create a mental cognitive map in a testing room.

During four of Henry's visits to the CRC from 1977 to 1983, we assessed his spatial ability using a route-finding task. The test's goal was to document his capacity to follow a route on a handheld map when walking from one landmark to another. Testing took place in a specially outfitted room at the CRC. Embedded in the tan wall-to-wall carpeting were nine red circles about six inches in diameter, in a three-by-three array. Henry held a large map that depicted the nine red circles on the floor as black dots. A path from dot to dot was drawn in heavy black lines, with a circle around the starting point and an arrowhead at the end. The test consisted of fifteen such maps. The letter **N**, designating north, was marked on each map, and a large red **N** was fixed on a wall of the room. Henry's task was to walk from dot to dot along the path that corresponded to the one on the map. He was not allowed to turn the map, so as he walked, the map was not always in the same orientation as the room. North was always at the top of the map, but when he turned, the **N** on the wall could be to his left, his right, in front of him, or behind him. As a result, he had to make a series of mental translations from the coordinate system of the map to the directional correlates of the room. Henry walked patiently from dot to dot, but was usually unable to follow the path indicated on the map, and his performance did not improve with repeated testing using the same maps (see Fig. 6).

The network of brain circuits that enabled him to draw the floor plan could not support his performance on this laboratory test. He needed his hippocampi for map reading. Without these structures, Henry could not fathom the relation between the start and finish points on the CRC floor plan, and could not reconcile the changing position of his body with the static coordinates of the room.

Henry's successful acquisition of a cognitive map of his house and his failure on the route-finding task seem contradictory. But the tasks themselves are fundamentally different. Through countless hours of *practice*, Henry slowly learned the geography of his house, without awareness and without consciously referring back to his declarative memory store. While the map test at the CRC was not a memory task, it did require Henry to form an instantaneous cognitive map, a task he could not perform without his hippocampi.

During the 1990s, as scientific knowledge grew with respect to how the brain processes complex thought, my colleagues and I continued to ask whether the preserved areas in Henry's brain could support new learning about the physical world around him. In 1998, a young neuroscientist at the University of Arizona studied patients who had received small lesions to their right hippocampus or right parahippocampal cortex to alleviate their epilepsy. Patients with right hippocampal lesions were unimpaired on a spatial memory task, but those with damage to the right parahippocampal cortex exhibited severe impairment, suggesting that the parahippocampal cortex is vital for spatial memory. Henry's right parahippocampal cortex was partially intact, so we wondered whether this part of his brain could support new place learning—the ability to navigate to a hidden target. To test this hypothesis, the Arizona researcher traveled to Boston to give Henry a simple spatial memory task, which he performed over nine test days during two visits to my lab in 1998.

On the first trial, the researcher told Henry that there was a sensor hidden under a small carpet, but she did not show him where it was. The testing room was filled with objects—desks, chairs, shelves, and a door—which Henry could use to get his bearings in relation to the environment. On this learning task, Henry had to find the invisible sensor by chance, remember its location, and then find it again from memory. Before each trial, the researcher triggered the sensor by stepping on it while Henry looked the other way. She asked him to find the place under the carpet that would produce the sound when he stepped on it. Because the sound triggered by the sensor came from a distant speaker, Henry could not rely on the sound to find the sensor's location. He was highly motivated in his search, even though he had to use his walker to navigate. On the first trial, he found the target. On fifty-four percent of subsequent trials, he walked directly toward the center of the carpet, and from there, on eighty percent of the trials, he took a direct path to the hidden sensor.[17]

Henry's ability to locate the sensor was remarkable, given his severe amnesia and inability to explicitly recollect the testing episode or having heard the sound. His achievement on this task underscores the role of the parahippocampal cortex, of which three-quarters of an inch remained in his brain, in spatial memory. We know that he was not relying on his short-term or working memory to perform the task because more than sixty percent of his hits—finding the sensor—occurred a day after the first test session. His capacity to locate the sensor indicated that he was capable of limited long-term memory formation, and that structures beyond the hippocampus could support navigation.[18]

But because Henry was unable to consciously recollect any details of the testing episodes, we concluded that learning the sensor's location was nondeclarative—learning that took place independently of the medial temporal lobe. Henry repeatedly found the sensor because of his implicit, nondeclarative knowledge of the target location. Whether his preserved parahippocampal cortex was solely responsible for his successful place learning was unclear because several other intact structures, like the striatum (located under the frontal lobes), could also have mediated this learning.

Although it took decades to discover the detailed anatomy of Henry's removal, one anatomical fact that we could hang our hats on from the beginning was that a large chunk of hippocampal tissue was missing on both sides of his brain. The maze experiments that Milner and I conducted with Henry helped establish the importance of the hippocampus for spatial learning. The later finding that he also performed poorly on the map-reading test, which did not require memory, indicated that his spatial deficit went beyond learning. Without his hippocampus, he could not process complex spatial information efficiently. He lacked the capacity to create a cognitive map in the usual sense. Other test results, however, highlighted an exception to the cognitive map theory and suggested a fractionation of spatial memory. Henry's unexpected drawings of the house where he lived when neither hippocampus was in working condition means that other brain areas took over the job of encoding and storing that rich spatial information. An indication of a specific brain structure that Henry likely engaged when drawing his floor plan came from the spatial memory task on which he was able to find the sensor hidden under the rug. Previous work had shown that this task depended on the parahippocampal gyrus, part of which remained in Henry's brain on both sides. So, on rare occasions, he somehow compensated for the devastating effect of his hippocampal damage by mobilizing preserved brain structures and networks.

A basic requirement for memory formation is intact perception. Henry passed this hurdle for sight, hearing, and touch, allowing my colleagues and me to test his learning and memory across these sensory modalities. His amnesia permeated all kinds of declarative memory without regard to the sensory portal that ushered in the to-be-remembered information. We consistently documented Henry's disorder with a wide range of test stimuli—words, stories, faces, pictures, scenes, mazes, puzzles, and more. Observations of his everyday behavior supplemented the knowledge that we gleaned from extensive formal testing in the laboratory, giving us a full picture of his postsurgery life.

Six

"An Argument with Myself"

Henry seldom shared his introspections with anyone, so, for the most part, we had to infer his emotional life from observing his behavior. During our conversations with him, he seemed happy and content; he smiled often and rarely complained. You might imagine that if you were in his shoes, you would be habitually anxious, concerned that your behavior had been improper and fearful of what tomorrow would bring. But no one would describe Henry as a nervous or worried man. It is possible that his operation, which excised part of his emotional brain, protected him from the scary realities of his life. Still, he did have occasional dark episodes when he could become frustrated, sad, aggressive, or uneasy. Typically, these negative emotions would dissipate as soon as he was distracted.

At the time of Henry's first visit to the CRC in 1966, his mother was in the Hartford Hospital after undergoing minor surgery. His father packed Henry's clothes and brought him to Scoville's office in Hartford, where Teuber had offered to pick him up and take him to Cambridge. Henry and his father had visited his mother in the hospital that morning, and by the time he met up with Teuber, Henry had only a vague sense that something was wrong with her. When Teuber asked who packed his bag for him, he answered, "Seems like it was my mother. But then that's what

I'm not sure about. If there is something wrong with my mother, then it could have been my father." During the trip to Cambridge, Teuber repeatedly explained to Henry where his mother was and that she was fine, but Henry had a lingering feeling of uneasiness about his parents, wondering whether all was well. As he settled into his room at the CRC, his anxiety dissolved. We told him that he could call home, but he no longer knew why he would want to do so. The next afternoon, however, Henry told a nurse that he thought his mother was in the hospital or had heart trouble. He had regained some inkling overnight that his mother was ill.

At that time, it was unclear what would account for the recovery of this memory; we speculated that Henry was simply less tired than the day before. Since then, however, numerous experiments in animals and humans have demonstrated that sleep sometimes improves memory consolidation. During sleep, memories may be reactivated and replayed, making them stronger and less susceptible to disruption. Different kinds of memory are enhanced by different stages of sleep, which in turn recruit different brain structures. For example, conscious, declarative memory performance benefits from deep (slow-wave) sleep, while unconscious, nondeclarative memory is enhanced by light (rapid eye movement, REM) sleep. Researchers have also found that REM sleep improves memory for emotional (especially negative) information more than nonemotional. According to the CRC nurse, he "slept fairly well" during his first night, and activation in preserved brain areas, including emotion circuits, may have strengthened a fragmentary memory of his mother's illness so that it surfaced the next day.[1]

The hospitalization of Henry's mother had dual representation in his brain, consisting of a factual element and an emotional element. He quickly lost the factual content—*Mother is in the hospital having minor surgery*—but the vague emotional content—*something is wrong*—lingered for days. Without a functioning hippocampal circuit, Henry could not preserve the facts about his mother's hospital stay in his long-term memory; but a larger network of brain areas, the *limbic system* and its connections, helped maintain his anxiety. The emotional component of the situation had privileged access and processing that helped establish an affective memory trace. Limbic is an anatomical term that means margin, referring in this case to a continuous band of cortical and subcortical structures nearest the boundary of the cortical covering. In 1877, this ring of cortex was originally believed to be concerned with the sense of smell, but a new proposal in 1937 described it as an anatomical basis for emotional behavior (see Fig. 7).[2]

In the 1937 version of the circuit, information traveled from one brain area to another in a loop—the hippocampal formation to the mammillary bodies of the hypothalamus to the anterior thalamus to the cingulate cortex to the parahippocampal gyrus and back to the hippocampal formation. In 1952, another researcher added the amygdala to the circuitry. The hippocampus is no longer believed to mediate emotions, whereas the amygdala is seen as the hub of emotional responses. This complex structure receives information from all the senses and from areas that process feelings of well-being and distress. The amygdala also sends information back to many of the same areas, creating vast networks specialized for emotional perception, expression, and memory. A mounting number of studies reject the idea that different emotions map one to one onto specific brain areas, recognizing instead that each emotional response evolves from a collaboration among many brain areas. By this view, the brain stores individual emotional experiences, for all kinds of emotion, by recruiting a wide range of brain structures in the limbic system and beyond—networks that support basic cognitive operations, both emotional and non-emotional. Henry's brain was occasionally able to create such networks.[3]

The removal of Henry's amygdala and hippocampus caused a malfunction in his basic limbic circuitry, so it was reasonable to expect that his ability to process emotions might be altered. But we learned from our earliest studies of Henry that he could experience a range of emotions. During his first visit to MIT in 1966, a CRC nurse woke Henry up each day at 4:00 a.m. to take his vital signs. On these occasions, she chatted with him briefly and then made careful notes in his chart. On eight of the sixteen nights of his stay, he inquired where his parents were, and whether they were OK. With parts of his distributed limbic system still functioning, he could feel anxious about his parents—and was doomed to re-experience this emotion.

Memory can be a burden: it forces us to revisit unpleasant events from the past. But without his memory, Henry could never properly mourn or process the losses that are an inevitable part of life. He did not remember that a favorite uncle had died in 1950. According to his mother, he became distraught each time he heard the news. As this emotion gradually faded with the passage of time, he would occasionally ask when his uncle would visit again.

In 1966, the same year as his first visit to the CRC, Henry suffered a tremendous loss. In December, when Henry was forty, his father died of emphysema at Saint Francis Hospital in Hartford. Mrs. Molaison told me that Henry was quite depressed after her husband's death but did not consciously grasp that his father was gone unless someone reminded him. She told us that at one point, Henry became angry and rushed out of the house when he discovered that some of his prized guns were missing, claimed by an uncle after his father's death. When the uncle learned that Henry was upset, he returned the guns, and Henry's anger subsided. He was disturbed by the disappearance of the gun collection because it was a focal point of his world. The guns had been on display in his room every day since his youth, so naturally their absence was noticeable and distressing. They also represented an emotional bond with his father and were prized possessions in their own right.

For at least four years, Henry was unable to articulate the fact that his father had died. Seven months after Mr. Molaison's death, Mrs. Molaison asked us not to tell Henry that his father was dead because he might react as if hearing the sad news for the very first time. My inclination was the same: I did not question him explicitly about his father because I knew it would upset him. But in August 1968, when I was testing him, he talked about his father in the past tense, so over time, his brain may have absorbed the painful fact into unconscious memory traces that stored it. Still, he would have moments of uncertainty. Without a functioning hippocampus and amygdala, Henry did not form long-term emotional memories; instead, he used what he had at his disposal—numerous interconnected cortical areas that stored preoperative memories of his father, his home life growing up, and the concept of death. Over time, he gradually connected the dots and understood on some level that his father was gone for good.

After the death of her husband in 1966, Mrs. Molaison continued to be Henry's sole caregiver. In the hope that increased activity would lift his spirits, she secured a place for him at the Hartford Regional Center, a workplace for mentally disabled people. There, he performed simple, repetitive jobs such as packaging multicolored rubber balloons into small bags, or attaching key chains to simple cardboard display stands. Every morning, Henry's neighbor Arthur Buckler—a short, rotund man in his sixties, who had an affectionate, sentimental attachment to Henry—drove him to the Regional Center. There, Buckler oversaw the grounds-maintenance crew and worked as a vocational teacher. He guided Henry through menial tasks like packaging balloons, instructing him to count the balloons and, when each bag had the right number, staple it closed. Henry did not suffer from low intelligence; in fact, his IQ was above average—120 in 1962. But even though his intelligence far exceeded the demands of this simple task, he had trouble with it because of his amnesia. He would become engrossed in inspecting the balloons one after the other and forget to stop at the right number. One day, however, Henry had an idea about a better way to do one of the jobs. A distant relative remembers Henry's contribution as "a mechanical thing that eliminated some steps." He shared his idea with the staff, and they adopted his suggestion. He must have been very proud of himself, if only for a moment.

Buckler later employed Henry as a general handyman at the Center, painting buildings, tidying up the machine shop and boiler room, and helping with outside maintenance work. Buckler would sometimes send Henry to the tool shop to fetch a hammer or wrench, only to find that Henry had forgotten his mission by the time he reached the shop. Buckler resorted to drawing a picture of the tool he needed on a piece of paper for Henry to take with him, which turned out to be a successful strategy.

In the spring of 1970, when Henry was working at the Regional Center, he experienced a breakdown. Mrs. Molaison noticed that Henry was much more nervous and irritable. When she talked to him, he would sometimes snap back brusquely instead of answering with his usual good-naturedness. One Sunday afternoon, he was behaving strangely, sitting around with his eyes closed, saying he wanted to be left alone. At one point, he jumped up and started pounding the door. Around five o'clock, he had a seizure, his body becoming rigid all over and his head lolling from side to side. After about ten minutes, the seizure ended, but instead of falling asleep after the seizure as he normally did, he started having one petit mal after another. Each time, he was briefly unresponsive and after, when his mother spoke to him, he said, "Stay out of my way," and slammed the bedroom door. It took him an hour and a half to get undressed for bed.

By the next morning, Henry's behavior was back to normal. He woke up and, as part of his usual ritual, asked his mother, "Now what am I doing today?" She answered that he was going to his job at the Hartford Regional Center, so he dressed and got in the car with Mr. Buckler. Because Henry's left hand was swollen and bruised, they stopped at Manchester Hospital to have it X-rayed. The films showed that he had broken his little finger, and it was put in a cast.

Two weeks later on a Tuesday morning, when Henry was packaging balloons at a table at the Regional Center, he went into a totally unexpected and unprecedented rage. He jumped up, shouted that someone had taken his balloons, and yelled that he had no memory, was no good to anyone, and was just in the way. He threatened to kill himself, and said he was going to hell and would take his mother with him. When others approached him, he kicked at them, and even flung a man across the room. He then turned to a wall and began to bang his head against it forcefully. A doctor arrived and injected Henry with a sedative, and after he calmed down, he was driven home to his mother.

Henry resumed work the next day without incident but seemed nervous. Mrs. Molaison called Teuber to report her son's alarming shifts in behavior. She believed the outbursts were some sort of seizure, and she was not afraid of his threats. She also wondered whether he was more upset because his memory was improving—or at least, she thought sometimes that his memory was better than it used to be. Perhaps, she reasoned, Henry was becoming more conscious of his situation and growing despondent with the realization that he was different from everyone else. She may have been correct: the constant repetition of memory failures at home and at the Regional Center possibly fostered despair and a sense of worthlessness in Henry. Over time, he gained the insight and accepted that he had a bad memory and that his condition was permanent. Mrs. Molaison worried that he would be sent to a mental hospital if he continued to have public tantrums, or that he would no longer be allowed to work at the Regional Center, which she believed gave him a sense of purpose.

Henry continued to have occasional tantrums, sometimes showing frustration at his inability to remember. Then in May 1970, he developed a new symptom, severe stomach pains, especially in the mornings, and one night, the sound of his moans woke his mother. She noticed that the pains seemed worse on weekdays when he was scheduled to go to the Regional Center, which made her wonder whether someone there was teasing him or giving him cause to be upset. She thought the pain might represent a dimly perceived anxiety he felt on the days when she told him he was going there. She also felt he was surlier in the mornings, except on his days off. These problems troubled Mrs. Molaison; she had no way of knowing the true cause of Henry's symptoms and did not know how to cope or help him.

The same month, Henry's mother called Teuber and explained what was happening. After Henry continued to complain of severe stomach pains over the next week, Teuber reassured her that he would arrange for Henry to have a thorough physical examination. Mrs. Molaison was worried that he was losing weight and said that the staff at the Regional Center thought his health was deteriorating.

My colleagues and I felt an obligation to ensure that Henry was looked after, even though we were researchers and not caregivers. Teuber consulted with Milner, and then with Scoville, who felt that he could not arrange for an examination in Hartford without creating substantial medical expenses for Henry's mother. In the end, they decided to bring Henry to Cambridge for an extended visit and medical evaluation at the CRC at no cost to the Molaisons. All of Henry's medical attention at MIT was free and placed no financial burden on the Molaison family.

Accompanied by his son Christopher, Teuber drove to East Hartford to pick up Henry. They pulled up to the one-story house on Crescent Drive, a small building in cream-colored wood with white trim, surrounded by a simple yard with a lawn and several trees. Inside, they found Henry neatly dressed, with a small suitcase packed for what would be his third visit to the CRC, a planned stay of three weeks. Mrs. Molaison thanked Teuber for his assistance, and remarked that this would be her first vacation from her role as Henry's caretaker in nearly twenty years. She was looking forward to visiting people without Henry and to going out in the evening.

During those three weeks at the CRC, Henry underwent a thorough medical examination by MIT physicians, but his stomach pains seemed to have subsided. Henry remained calm but disoriented throughout his stay. One evening, Teuber found him sitting in his darkened room, a crossword puzzle nearby. He asked Henry whether he was feeling any physical discomfort. "Well, mentally, I'm uncomfortable," he said, "to be so much trouble to everybody—not to remember." He searched for the right words. "And I keep debating with myself if I said anything that I shouldn't have, or done something that I shouldn't have done." Whenever Henry was struggling to retrieve a memory, he would say, "I'm having an argument with myself." It was a constant refrain. Teuber reassured him and said he would call his mother that night. In an unusual moment, Henry turned to the subject of his father. "There I am having a debate with myself—about my dad," Henry said. "You see, I am not easy in my mind. On the one side I think he has been called—he's gone—but on the other I think he's alive." He began to tremble. "I can't figure it out."

Henry was not an anxious man, but the conversation with Teuber revived his sadness and uncertainty surrounding the death of his father. He could not hold on to the fact that his father had died long enough to come to terms with his death. He had no memory of saying goodbye to his father, attending his funeral, visiting his grave, or being comforted by the love and sympathy of his family and friends. The trembling that Teuber observed was a physical expression of Henry's emotional state.

After three weeks at the CRC, it was time for Henry to return home. He did not complain about abdominal pain during the entire visit, suggesting that the problems that prompted his CRC admission were stress related, likely stemming from his activities at the Regional Center. Henry was smoking a pack of cigarettes a day, and X-rays detected lung disease not visible in his 1968 X-rays. Teuber called Mrs. Molaison to discuss Henry's return trip and then passed the phone to Henry. He was visibly moved, his voice nearly breaking as he told his mother it was good to hear her. Teuber drove Henry back to his house in East Hartford. Mrs. Molaison came to the door when the car pulled into the driveway and remarked how well Henry looked, much better than when he left. The two of them embraced without speaking for a few moments, as Henry stroked her cheeks and shoulders.

Clearly, Henry could feel and communicate emotions—both positive and negative—despite missing almost all of his amygdala, one of the key structures underlying emotions. In formal testing, he could judge the emotion in pictures of faces, for instance, indicating whether an expression was happy or sad. He was usually on an even keel, but on rare occasions would become very angry. These aggressive episodes were short-lived responses to his frustration at not remembering important information and to fellow patients who irritated him. Henry was not a violent man. On the contrary, he was mild mannered, friendly, and patient, and his behavior in social situations was exemplary. At the CRC, he was always docile and friendly.

The science of emotion explains why Henry could experience and exhibit positive and negative moods. Emotions cover a wide range of experiences. In 1969, psychologist Paul Ekman proposed that people across all cultures experience six basic emotions: sadness, happiness, anger, fear, disgust, and surprise. Various combinations of these core emotions give rise to many others, such as affection, hope, empathy, ambivalence, outrage, and shame. Feelings vary with respect to two distinct variables: the extent to which they are pleasant or unpleasant, and the extent to which they are arousing or calming. Underlying your conscious experience of emotions are increases in your heart rate, blood pressure, breathing, blood glucose, and stress hormones. An accompanying increase in the flow of blood to your body and brain prepares you for actions that enable you to express your emotional state. You are ready to run away, fight, or hug, depending on the situation. Your brain regulates all the biological variables, and the brain circuit that is activated varies with the nature of the emotion being generated.[4]

In 1970, on the drive up for his third visit to the CRC, Henry witnessed an unusual incident that gave us additional insight into the nature of his emotional memory. When Teuber picked up Henry in Hartford, it was raining heavily. Route 15 was awash with water and mud as Teuber, his son Christopher, and Henry began driving north toward Boston. Teuber was in the right-hand lane when a light brown Impala ahead suddenly spun out of control, ramming into a steep bank on the right side of the road. The car teetered on its left wheels and then settled back on all four wheels. The front frame was bent, and the deflating rear wheels hissed loudly over the sound of the rain.

Teuber stopped his car a few feet behind the wreck, which protruded into the right-hand lane. He told Henry to stay in the car and left to check on the passengers—a twenty-year-old girl and her mother, both upset but unhurt. The mother, a large woman, began crying hysterically, while another car veered around the wreck and stopped. A young man jumped out of the car, and he and Teuber led the women away from traffic. Teuber returned to the car for a raincoat, but decided he was too drenched for it to matter. After looking at the car, they decided it could still be driven on the flat tires. Teuber blocked traffic while the young man drove the wrecked car off the embankment and parked it in the emergency lane at the side of the road.

With everyone out of danger, Teuber returned to his car and drove away. For several minutes, Henry spoke worriedly with Christopher about the possible causes of the accident. Their conversation then drifted to the continuing downpour. About fifteen minutes after leaving the site of the accident, they passed a police car with lights flashing, parked in a turnout at the side of the road, behind a blue station wagon with a red trailer attached. Henry remarked that the police car must have been protecting the car with the trailer so that no one would pull into the turnout behind it. After a pause, Teuber asked Henry, "Why am I all wet?"

"Well, you went out to offer help in an accident, when a car went off the road," Henry answered.

"What kind of car?"

"A station wagon—no, a station wagon and a truck."

Teuber asked Henry the color of the wrecked car.

"There I am having a debate with myself," Henry answered. "The station wagon that went off the road—it was lying on one side—was blue. But then it comes to mind—brown." A couple of minutes later, Henry said that a state trooper had been at the accident redirecting traffic. About twenty minutes later, Teuber asked Henry again why he was wet.

"Because you got out—to inquire about the way."

The accident, an emotional event that had made a vivid impression on Henry, gradually became intermingled with new information—a police car, flashing lights, a different vehicle at the side of the road—that pushed out the old memory. Over a very short time, the memory seemed to have vanished completely. Still, the intense excitement of the incident left an unusually clear impression in Henry's mind.

After arriving at the CRC, when asked about the drive, Henry said that there had been a lot of traffic, and they went through a detour, but nothing else had been amiss. That night, however, Teuber again asked Henry if he remembered anything about the trip that day. Henry said no.

"Did I get wet?" Teuber asked.

"Yes, you got wet when you went out after the accident."

"What accident?"

"When the car spun around, and went part of the way up the bank—a girl was in it. You went out into the rain to see if anyone was hurt."

"Was there just one person in the car?"

"No, another one—another lady—a fat one."

This story illustrates several principles central to the nature of memory function. On this trip, Henry encountered two exciting episodes that caught his interest and raised his level of arousal. When we pay careful attention to an event and are excited by it, our memory for that experience will be enhanced. The circuits in Henry's brain devoted to attention and emotion were firing away during this episode, so he was able to encode details of the people, vehicles, and action. Just at the time when Henry might still have been reliving and rehearsing these details, the three travelers encountered a police car and trailer, which caught Henry's attention. Henry turned his focus to this second event and absorbed all that was going on there. In doing so, however, he interfered with the processing of the first event and lost some pieces of information.

Interference is a major cause of forgetting for all of us. In this case, the new encounter involving the police car, station wagon, and trailer competed with information from the accident and interfered with the maintenance of the older information. Farther down the road, Teuber cued Henry's memory of the accident by asking why he was all wet, and Henry's reply blended details of the two episodes, indicating that both memory traces were fragile and incomplete.

In healthy brains, such delicate memory traces become more robust over time through a process called consolidation. Henry was unable to consolidate new information because that activity required interactions between the hippocampus and the cortex—impossible in his brain. When consolidation occurs, emotional situations enjoy privileged processing in memory circuits, resulting in memories that are relatively hard to erase. By the time Henry reached the CRC, he had no recollection of the trip's excitement, but that night, Teuber again cued him by asking whether he got wet. As before, this tip helped Henry to recall the action, the girl, and the lady. By then, he had had time to rest, have dinner, and re-experience whatever fragments of his trip he had been able to hold onto. The mechanism by which he did this could not have been the same as in a healthy brain, but, remarkably, he successfully engaged other circuits to establish temporary traces. At the time, we did not know about the remaining medial temporal-lobe tissue around the hippocampus, which together with preserved emotional memory areas, likely supported Henry's brief recollection.[5]

Until the early 1980s, our information about Henry's emotional state came from people who witnessed and documented his emotional behavior. Around that time, my colleagues and I decided to take a more objective and broader look at Henry's personality. Because his operation damaged his amygdala, it was appropriate to formally examine his emotional state. We had not done this evaluation previously because in the 1960s and 1970s, many neuroscientists, including members of my lab, shunned topics that belonged in the realm of clinical psychology and psychiatry.

We gave Henry a range of standardized personality tests. The results revealed that his level of emotion was somewhat blunted, but he was still capable of displaying a range of emotions. We also determined that with respect to self-care, he was somewhat negligent and required supervision. For example, someone had to prompt him to shave and take a shower. Measures of his personality and motivation indicated that he was socially interactive but lacked initiative. Importantly, the tests showed no evidence of anxiety, major depression, or psychosis. Healthy people experience grief, sadness, and frustration, not unlike Henry's emotions when his father died and when he could not remember things. On rare occasions, he did become extremely angry, but this kind of reaction is not unheard of in people faced with serious handicaps.

In 1984, I asked a psychiatrist, George Murray, to evaluate Henry. Murray noted that Henry "was always smiling, and had a relatively warm interaction with me." Henry did not know whether his appetite was good or bad but smiled when he reported that he did not like liver.

When Murray asked him whether he was sleeping normally, he replied, "I guess so." He said that he did not think of death and, to his knowledge, did not cry. When asked if he felt helpless, he said, "Yes and no," and if he felt hopeless, with a wide smile said, "Yes, and mostly no." When asked if he felt worthless, Henry smiled again and said, "This could be the same as hopeless." The preceding question had lingered in his short-term memory. When Murray asked him whether he liked himself, he once more had a cautious smile and said, "Yes and no—I can't be a brain surgeon." (Over the years, a recurring theme in Henry's conversations was that he had wanted to be a brain surgeon.) Murray concluded that Henry "does not have any depression. This does not mean that he cannot feel sad on occasion."

Murray continued to probe Henry's emotional life with additional questions about his parents and taste in music. They laughed together over their mutual dislike of "jive." Then Murray moved into the area of sex. He asked Henry if he knew what an erection was, and he said "a building." Then Murray said, "well, let me use some other language" and asked him what a "hard-on" was. Henry, without smiling, frowning, or any change in facial muscular said, "a man gets it, below the belt." Henry knew that men have penises and women do not, and he described how babies are conceived. During this line of questioning, Henry had no facial responses to Murray's questions and said that he did not have any sexual desire. Murray described him as *asexual*—having no libido. (Buckler, Henry's boss, had described Henry as a perfect gentleman who "never as much as looks at the girls at the Center.")

When my colleagues and I interacted with Henry, he was always friendly but passive. He had an excellent sense of humor that occasionally popped up in everyday conversation. For example, one day in 1984, a neurologist in our lab walked with Henry out of a testing room into the hall. As the door closed behind them, the neurologist wondered aloud whether he had left his keys inside the room. Henry replied, "At least you'll know where to find them!"

Henry's inherent easygoing and generous nature was apparent in the supreme patience he exhibited in undertaking all our tests. Of course, he had no long-term memory for the testing episodes, so each one was a new experience for him, and he never seemed bored. Once, when talking with one of the members of our lab, he summed up his testing experiences this way: "It's a funny thing—you just live and learn. I'm living, and you're learning."

Seven

Encode, Store, Retrieve

In 1972, I visited Mrs. Molaison and Henry when the Watergate scandal was dominating the news and asked him what Watergate meant to him.

"Well, I think of a prison right off, and I think of a riot in Watergate Prison," he replied.

"Have you heard anything on the news recently about riots or about Watergate?" I asked.

"No. Then I think of an investigation into it."

"That's right," I said, encouragingly.

"But that, uh, I can't put my finger more on it, I guess."

"Have you ever heard the name John Dean?"

"Well, I think of an assassin right off, but then, after I said that, after I said assassin, I thought of, uh, a leader, you know, labor leader or worker that was killed or injured. That's what I think of."

"You read all about them in the papers and everything," Mrs. Molaison interjected.

John Dean was Nixon's White House counsel. Henry had been exposed to the extensive coverage of the Watergate break-in, but like a computer with a faulty hard drive, his brain was unable to store and retrieve this information.

The modern study of the human brain owes much to advances in computer science. Our search for the underlying cognitive operations of long-term memory is now grounded in *information theory*, an idea introduced in 1948 by Claude Shannon, an engineer at New Jersey's Bell Telephone Laboratories. Shannon introduced this idea in his mathematical theory of communication, integrating knowledge from applied mathematics, electrical engineering, and cryptography to describe the transmission of information as a statistical process, and coining the term *bit* to refer to the most fundamental unit of information. In the early 1950s, the cognitive psychologist George A. Miller introduced information theory to the study of natural-language processing, thus integrating Shannon's ideas into the field of psychology.[1]

Conceptualizing learning and memory in terms of information processing was a key advance, allowing researchers to divide memory into three stages of development, akin to computer processes. The first stage is *encoding* information by turning sensory inputs from the world into representations in the brain. The second stage is *storing* those representations so they can be extracted later. The third stage is *retrieving* the stored memories when they are needed. Researchers now design memory experiments to examine each of the three stages separately, and to watch how they interact.

Scientists divided the underlying information-processing stream into these discrete stages to make the scientific study of memory tractable. This artificial division is a simplification, but a necessary one: it allows researchers to describe in detail the numerous processes within each stage. In reality, encoding, storage, and retrieval occur constantly and simultaneously. Understanding the constituent parts of memory formation is essential to assembling them into a comprehensive theory.

Henry had no trouble encoding information. When I would ask him whether he wanted milk in his tea, he could register my question in his short-term memory and reply that he never took milk but did take sugar. Henry's problem was with the last two stages of information processing—storage and retrieval of new information. If I distracted him by changing the topic of conversation and then asked what we had talked about earlier, he would not know. The stimuli his brain received could be held briefly, but they could not be squirreled away and revisited later.

Beginning with the publication of Scoville and Milner's paper in 1957, Henry's case helped to launch decades of research that dissected the cognitive and neural processes within each of the three stages of memory formation. Just as important, Henry's case illustrated the fractionation of memory—the idea that our brains constantly juggle different kinds of short-term and long-term memory processes, each mediated by a separate, specialized memory circuit. Milner's landmark discovery, showing that some of Henry's long-term memory processes were disrupted while others were not, led to the important theoretical distinction between *declarative* or explicit memory—seriously impaired in Henry—and *nondeclarative* or implicit memory—intact in Henry.[2]

Declarative memory, rooted in the medial temporal lobes, refers to the type of memory we invoke when in everyday conversation we say, "I remember" or "I forget." This kind of memory includes the capacity to recollect consciously two kinds of information: *episodic knowledge*—the recollection of specific experiences we were part of in the past—and *semantic knowledge*—general knowledge, such as information we gather about people, places, language, maps, and concepts, not linked to a particular learning event. In many ways, declarative memory is the backbone of everyday life, enabling us to acquire the knowledge we need to pursue goals and dreams and to function as independent people.

Henry lived for fifty-five years without acquiring any new declarative memories. He could not tell us about exact incidents, such as what he had for breakfast, what tests he had taken the day before, or how he celebrated his last birthday. He also could not learn new vocabulary words, the name of the current president, or the faces of people he encountered at the CRC. On memory tests, his scores were no better than if he had guessed at the answers. The structures removed from Henry's brain were dedicated to declarative memory. His surgery, however, left intact other circuits that supported his nondeclarative memory, so he could learn new motor skills and acquire conditioned responses.

Research that stemmed from Henry's case shed light on the fundamental processes that underlie the encoding, storage, and retrieval of episodic knowledge. Over the last fifty-five years, scientists have made great progress in characterizing these three stages of processing. In the 1990s, these investigations were fueled by the advent of brain-imaging tools such as positron emission tomography (PET) and functional MRI. These technologies made it possible for researchers to examine, for the first time, brain activity for each stage of processing separately.

After the discovery that conscious remembering depends on mechanisms in the hippocampus and its close neighbor, the parahippocampal gyrus, scientists began to tackle basic questions in the psychology and biology of episodic learning: What specific cognitive operations contribute to long-term memory of a single event? What are the complex workings within the hippocampus and parahippocampal gyrus? What is the role of the cerebral cortex in long-term memory? What cognitive processes and corresponding brain circuits mediate how well we encode, store, and retrieve episodes, and how much we forget?

Simply registering a sensory event—seeing, hearing, smelling, touching, or tasting—does not ensure that learning will happen. How well we remember events and facts depends greatly on how effectively they are encoded initially. The likelihood that we will remember a name, face, date, address, directions to a party, or anything else is related to the richness of the representation. Researchers call this the *depth-of-processing effect.*

Psychologists Fergus Craik and Robert Lockhart first described this effect in the early 1970s after conducting a series of experiments to study how deeply their subjects processed information. The researchers persuasively argued that when the brain receives information, it can process that information to different depths. Craik and Lockhart gave their subjects short words such as *speech* and *daisy* as test stimuli, allowing participants to view each word briefly, and then asking them a question about each word. By using three kinds of questions, the researchers hoped to generate different levels of processing—shallow, intermediate, and deep.[3]

Envision the printed word *TRAIN*. Craik and Lockhart encouraged shallow processing through questions about the physical structure of the word (*Is the word in lowercase?*), intermediate processing through questions about the rhyming characteristics of the word (*Does the word rhyme with* brain?), and deep processing through questions about the meaning of the word (*Is the word a mode of travel?*). After participants encoded a list of words in this way, there was a brief pause, followed by a surprise memory test to see which words they recalled. The experiments showed that subjects remembered best those words encoded by processing their meaning, followed by words encoded by their rhyming characteristics, and then by words encoded by their physical structure. Overall, participants' retention of the words depended on how elaborately and descriptively they thought about the words as they encoded them. Thus, Craik and Lockhart illustrated that deep processing produces stronger memories than shallow processing.[4]

In 1981, we became curious to see whether Henry would show the depth-of-processing effect. We designed a depth-of-processing test in which we helped him think hard about the meanings of words to enhance his ability to recognize those words later. Would he be more likely to recognize words he processed deeply than words he processed superficially? The stimuli for Henry's test were thirty common nouns such as *hat, flame,* and *map*. For the encoding task, the examiner played an audiotape. Henry first heard a word such as *hat*, and then one of three kinds of questions, which he answered "Yes" or "No." For example, "Does a woman say the word?" targeted the physical (shallow) level. "Does the word rhyme with *glass?*" targeted the phonological (intermediate) level. "Is the word a type of clothing?" targeted the semantic (deep) level.[5]

After the encoding phase, Henry took an unexpected memory test to see whether he recognized the words he had just encoded. The examiner read him three words, asking him to choose the one he had heard before and encouraging him to guess if he was unsure. Henry performed the depth-of-processing task on two occasions. Just by chance, he should have gotten ten out of thirty correct; and in two separate test sessions, his score was no better than chance—twelve in 1980, and ten in 1982.

His overall performance was deficient; to answer our original question, he did not show the depth-of-processing effect.[6]

We now understand that Henry did poorly not just because his hippocampus was damaged, but also because he lacked the critical connections and interactions that occur between medial temporal-lobe structures—where information is processed initially—and the cortical areas specialized for storing representations of words and other information. While the hippocampus makes a vital contribution to encoding, the cortex plays an equally important role. Functional MRI studies, which allow us to look at brain activity during task performance, show convincingly that activation in the cortex is greater during deep processing than during shallow. The cortex, however, cannot perform the job of encoding by itself. Henry's senses could perceive words, pictures, sounds, and touch, and could deliver this information to his brain where his cortex registered it. But beyond that stage, his ability to store that information was so deficient that deeper processing did not help. Although he could receive and understand incoming sensory information normally, he was unable, even with added elaboration, to form deep representations that would result in better memory.[7]

In general, the more deeply we characterize a name, face, date, address, or anything else, the better we remember it. This is true whether we extract the information from long-term memory without help—*free recall*—or it automatically jumps out at us when we consider several options—*recognition*. Imagine you want to search online for directions to a particular Italian restaurant. If your brain has a rich representation of the restaurant based on past visits, you will spontaneously recall its name and the town it is in and plug that information into your computer's search engine. On the other hand, if your representation of the restaurant is sparse because you have never eaten there and just drove by it once, you may not recall the specific name and will need to view a list of possible restaurants until you recognize the one you are looking for.

When we encode new information deeply, we are more likely to retrieve it later because we have linked it to a wealth of semantic information already stored throughout the temporal, parietal, and occipital cortices. *Elaborative rehearsal*, in which we mentally manipulate information and relate it to other facts we know, helps long-term memory much more than if we simply repeat the information. Students know the value of preparing for tests and exams by forming small study groups in which they ask one another practice questions drawn from lectures and readings. The ensuing discussions constitute elaborative rehearsal: they encourage deeper processing and more robust encoding of the material than if the students simply read over their notes silently in the library.[8]

Unlike our hypothetical students, Henry could not benefit from dynamic elaborative rehearsal. In 1985, however, Henry did use *simple rehearsal* to accomplish a feat that at first blush looked like long-term memory formation. A postdoctoral fellow in my lab wanted to test his appreciation of the passage of time. She told him that she was going to leave the room and that when she returned, she would ask him how long she had been gone. She left the room at 2:05 and returned at 2:17. When she asked Henry how long she was out of the room, he replied "Twelve minutes—got you there!" She was astonished, until she noticed the large clock on the wall and understood how Henry had come up with the correct answer. During her absence, he repeated 2:05 over and over to himself, keeping it foremost in his mind, and when she came back, he looked at the clock and saw that it was 2:17. He then harnessed his working memory capacities to do some simple arithmetic, subtracting 2:05 from 2:17. Henry could not remember things, but occasionally he could be quite clever in finding ways to compensate for his disorder.[9]

Elaborative rehearsal is not the only way to beef up memory. History is rich with examples of people using intricate techniques to recall information, creating mental representations and organizing them so that information is available later. In 1596, Matteo Ricci, an Italian Jesuit missionary and scholar working in China, wrote a short book entitled *Treatise on Mnemonic Arts*, laying out a technique for Chinese men to memorize the vast burden of knowledge they needed to pass challenging civil-service examinations. Ricci's technique, based on a medieval

European idea, centered around a "memory palace": an imposing edifice with a reception hall and many rooms that contained lively, complex images, such as emotion-provoking paintings, located at different points. Because items that have emotional content are more memorable than neutral items, the trick is to create an outrageous or emotional association between each piece of information to be remembered and an object in the room. In doing so, we form vivid mental associations. A modern term for Ricci's memory technique is the *method of loci*: picturing a familiar route in the mind's eye, placing distinct landmarks along that route, and associating different to-be-remembered material with each one.[10]

If you want to create a memory palace, choose a familiar landmark—an office building, a nearby grocery store, or your house. For example, let's say that you are trying to memorize a toast you will deliver to the bride at her wedding reception. You have a specific set of stories you want to remember—about soccer in elementary school, gymnastics in middle school, travel to France in high school, acquiring a dog in college, and later meeting the groom. Pick out a familiar location to become your memory palace—for instance, your local supermarket. Insert cues to the anecdotes in your speech into the orderly arrangement of the foods. You place an eye-catching reminder on the door of the market, and then progress, in succession, to the fruit, vegetable, meat, and frozen foods departments. As you approach the entrance, you picture an enormous soccer ball filling the glass door, with the bride and her best friend at age seven in their soccer uniforms, perched on top of the ball holding hands. Moving on to the fruit department, you imagine the bride's gymnastics team doing handstands on the watermelons, and in the vegetables, a giant Eiffel Tower atop asparagus spears. In the meat section, place a full-size Siberian Husky inside the showcase with a five-pound steak in his mouth, and in the frozen foods aisle, visualize the groom inside the freezer, on one knee, holding a gigantic bag of onion rings. Once you have constructed these images and fixed them in your mind, you can embark on a mental excursion through the supermarket, with your mind's eye traveling from one memory image to another. As you deliver your speech, you can retrieve, in a specific order, the memories and anecdotes you have stored there. People of all ages can take advantage of memory-enhancing tricks like this one.

Many people who enroll in high-level memory competitions use the method of loci. For instance, every Pi Day—March 14, or 3.14—Princeton University hosts gatherings of people to see who can recite the greatest number of decimal places in this mathematical constant. In 2009, researchers from several universities collaborated on a functional MRI study to identify the brain areas that support the ability to memorize pi to a staggering number of decimal places. During such an experiment, participants lie on their backs inside an MRI scanner and perform a behavioral task. As specific brain areas are activated, their oxygen consumption increases. The brain detects this increased use of oxygen, and orders more blood to bring oxygen to this part of the brain. The increase in oxygenated blood in this area changes its magnetic properties. We can detect these localized changes in the magnetic field using a powerful magnet, several thousand times the magnetic pull of the earth and guaranteed to render your credit cards useless if you bring them too close. In this way, we can map activation by identifying the brain circuit recruited for each particular capacity.

In the 2009 study, functional MRI allowed researchers to record the brain activity of a twenty-two-year-old engineering student as he recited the first 540 digits of pi. The student used the method of loci to remember the digits in order, and functional MRI scans conducted while he performed this feat showed that his retrieval processes elicited robust activation in particular areas in his prefrontal cortex. These areas are known to support working memory and attention, suggesting that the student engaged cognitive-control processes to rattle off the well-learned sequences of pi.[11]

To gain insight into how people acquire such vast amounts of information in the first place, the researchers asked the engineer to learn a novel series of one hundred random digits as he was being scanned. The result was impressive: after three six-minute scans, the student had encoded all one hundred digits in the correct order, using his own variation on the method of loci. During early stages of encoding, the functional MRI images showed greater engagement of cortical areas specialized for visual processing, emotion-related learning, motor planning, task scheduling, and working memory than during the retrieval of pi digits. These areas were activated because the task required a great deal of mental effort, and to succeed the student had to harness numerous resources, including visual processes in the back of his brain and cognitive control processes in the front.

The student explained that in his particular method of loci, he relied heavily on color, emotion, humor, vulgarity, and sexuality to build his memory palace—"the more emotional and gruesome the scene, the easier the recall." The researchers linked his consistent use of highly emotional images to a structural difference in his brain—increased volume of an area in his cingulate gyrus, part of the limbic system. This student was brilliant at memorizing groups of digits, but he did not necessarily possess a superior intellect or memory; he had worked hard and had developed highly effective cognitive-control circuits for retaining information. As one psychologist previously put it, "exceptional memorizers are made, not born." Phenomenal memory performance is within your reach, whenever you want to apply it. When you try to memorize names, numbers, words, pictures, and so on, you will remember the items better if your brain is optimally active during your initial exposure to the material.[12]

Encoding is the gateway to memory formation, with consolidation and storage following closely behind. Henry could encode the information presented to him and register it briefly, but then his processing broke down. He could not consolidate and store that information.

In 1995, when functional MRI was in its infancy, we had the opportunity to observe Henry's encoding processes in action. In this experiment, we asked him to look at pictures of scenes and indicate whether they were indoors or outdoors. These questions were intentionally easy. He answered them correctly, so we knew he was looking at and processing the pictures. The corresponding MRI images showed increased activity in his frontal lobes as he performed the picture-encoding task. Subsequent experiments in other labs with healthy research participants have expanded on this finding, and the results indicate that two separate areas in the left frontal lobe and an area in the right frontal lobe are normally active during encoding. Henry could fire up his frontal cortex to encode the objects he perceived, but after that, he was stopped in his tracks—the process of memory formation broke down, severely impeded by his inability to consolidate and store information.[13]

When the brain receives and encodes new information, the content must be processed further to make it available for future use. The initial transmissions are not immediately encased in long-term storage. The longer process by which memories become fixed, *consolidation*, is a lasting change that happens in individual neurons and their molecular components. Connections between adjacent cells become stronger or weaker in response to learning experiences. Henry's defunct hippocampus was unable to initiate and complete the active processes required for consolidation.

Two ambitious experimental psychologists at the University of Göttingen, Georg Elias Müller and Alfons Pilzecker, introduced the concept of consolidation in 1900. Ever since, scientists have struggled to understand the mechanisms by which the brain consolidates memories. This quest has inspired thousands of experiments in many species from insects to humans, and has fueled healthy controversy about how different kinds of memories are anchored in our brains.[14]

Müller and Pilzecker made the novel discovery that declarative learning, consciously retrieving facts and episodes, does not immediately lead to enduring memory. Instead, consolidation depends on changes in the brain that occur gradually over time. During this time, the newly learned material is susceptible to interference. The German researchers came to this conclusion after eight years of experiments performed on a small group of participants that included their students, colleagues, family members, their wives, and themselves. First, they created 2,210 nonsense syllables, assembled them into pairs such as *DAK-BAP*, and generated lists of six pairs. They tested one participant at a time, with the arduous training and testing taking place over twenty-four days.

During training, the participants read the list of six pairs aloud and tried to mentally associate the two nonsense syllables that formed each pair. Then came the memory test: the participants saw a cue—the first syllable of each pair, for example, *DAK*—and were asked to say the second syllable of the pair, *BAP*.[15]

The researchers' key insight came when they focused their analysis on the participants' *intrusion errors*. An intrusion error was scored if participants, while recalling the items in one list, inserted a nonsense syllable that was in an earlier list, thinking that it belonged to the current one. In this experiment, if participants had previously learned the syllable *JEK*, they might pair that with *DAK* instead of the correct answer, *BAP*. The psychologists reasoned that these intrusion errors resulted from the persistence of the just-learned material in the participants' recent memory. These errors were most numerous when the test was performed twenty seconds after training, when the brain was still encoding information, and became less frequent as the time between the training and the test increased from three to twelve minutes, the period during which the brain was consolidating information. Intrusion errors did not occur twenty-four hours later; by then, the subjects had successfully consolidated the information. The experiments revealed that consolidation is an active process that takes time. The associations were easily broken right after encoding but became stronger from minute to minute.[16]

Subsequent research with animals supported Müller and Pilzecker's hypothesis. In 1949, a physiological psychologist at Northwestern University trained groups of rats to avoid a grid that delivered a mild shock. They then administered electroconvulsive shock (ECS) to the rats' brains at various times after the end of each learning trial. The group that received ECS after twenty seconds was the most impaired, and the deficit became progressively less as the time between learning and ECS increased to forty seconds, one minute, four minutes, and fifteen minutes. The rats that received ECS after one hour or longer were unimpaired. The longer the interval between encoding and ECS, and the more time for consolidation, the better the memory. These results show that for a limited time after training, disrupting brain activity blocks the mechanisms underlying consolidation. In Henry's brain, the critical cellular events in the hippocampus and cortex that occur for minutes or hours after encoding never activated, and thus new declarative information could not be secured.[17]

From these and many other experiments, neuroscientists learned that the neural infrastructure of memories, the physical traces that exist within the brain, are initially frail and strengthen gradually. They can be disrupted by behavioral manipulations in the laboratory and by more direct insults to brain physiology in the form of drugs, alcohol, or head injury. A too-familiar example of the vulnerability of memory formation is prevalent in North American football. In the fall of 2012, a linebacker playing his first varsity high-school game went to tackle the opposing team's running back who was carrying the ball. They collided—helmet to helmet—and both players went down but soon got up and went back to their positions. After two more plays, the linebacker's teammates went to the sideline and told the coach that the injured kid was playing out of his position and, when told to move into his position, did not do so. He was immediately examined by the team's physical therapist who determined that his neurological status was normal, he was not nauseous, and he did not have a headache. The player's striking symptom was that he had no memory of the hit or anything that transpired after the hit. The blow to his head had derailed the consolidation of these fragile memory traces.

Most of the memory tests administered in clinics and laboratories assess declarative, episodic memory—the ability to form associations among words, elements in a story, or details in a picture. Before Henry, memory researchers did not know for sure which brain structures were responsible for establishing these connections. The crucial lesson Henry taught us is that the hippocampus is necessary for building associations. Without a functioning hippocampus, Henry found it impossible to form associations between familiar words; he could not link them in his memory. His long-term memory stores for new information were always empty.

Association, a basic concept in both animal and human learning, is the essence of episodic memory; it enables us to characterize a unique event (reading this chapter) by integrating its context in time (3 p.m.) and space (in the kitchen with light coming in the window). The context may be rich, including who else is in the room, whether music is playing, and specific thoughts about each sentence read.

In everyday life, associations develop and strengthen over time when particular items occur together repeatedly. When we move to a new neighborhood, we gradually get to know the people who live in our community—our neighbors and the people who work in the coffee shops, pharmacies, and restaurants that we frequent. Eventually, we get to know some of these people well, as we gather information about their personal lives bit by bit. We learn, for example, that the gentleman behind the espresso machine, who always asks about our dog, is a student who has been working on his degree for five years and who dreams of being a journalist, and that the older man at the convenience store who always has a cheery greeting lost his granddaughter to cancer. We experience what it is like to live in this neighborhood in the spring, summer, fall, and winter, and we store the sights, sounds, and smells that characterize the environment. Over time, our brains build up an elaborate representation of our neighborhood, in which lots of individual facts and events have become connected to one another. After we have lived there for a few years, we can provide a vivid, detailed picture of what the neighborhood is like.

In recent decades, thanks to contributions from thousands of labs in many countries, scientists have come to understand the cognitive processes and neural representations that support these kinds of associations. The cortical neighbors of the hippocampus, the parahippocampal areas, flood the hippocampus with complex perceptions, ideas, and contexts, and the hippocampus associates this wealth of information in three ways. *First*, the hippocampus links distinct objects with one another and with the time and place we encountered them—for example, all the objects and people we saw, the sounds we heard, and the aromas we smelled in our neighborhood coffee shop this morning at 7:55. *Second*, it links events in time to record the flow of experiences that comprise a unique episode—for example, the sequence of entering the coffee shop, getting in line, reading the menu, ordering a large cappuccino, waiting for the server to brew our drink, picking up our order, and rushing out the door to get to work. *Third*, it links many events and episodes in terms of their common features to form a network of relationships—for example, connecting this morning's coffee-shop memory to memories of meals in other coffee shops and restaurants that we frequent, thus composing our general knowledge of eating out.[18]

Each morning, when we encode the details of an experience in a coffee shop, this new learning reactivates many separate events from the past, resulting in an updated, rich associative representation that transcends the individual events. To establish this inclusive representation of eating out, we depend on cooperative interactions in our brains between our hippocampus and regions in the midbrain, a two-centimeter-long structure that connects the cortex and striatum to areas lower in the brain. Cross-episode integration, connecting separate experiences that have common characteristics, guides decision making in everyday life. (Should I go to the coffee shop that has the best cappuccino or to the one that has fabulous pastries?) This intricate cognitive and neural infrastructure was not available to Henry.[19]

When Milner tested Henry for the first time in 1955, she examined his ability to form word associations by reading aloud eight word pairs. Some of the pairs were considered easy to remember because their meanings were readily associated, while others were hard because the words were unrelated.

Metal–Iron (easy)
Baby–Cries (easy)
Crush–Dark (hard)
School–Grocery (hard)
Rose–Flower (easy)
Obey–Inch (hard)
Fruit–Apple (easy)
Cabbage–Pen (hard)

Five seconds after reading the list of word pairs, Milner asked Henry, "Do you remember what went with *Metal*? *Baby*? *Crush*?" and so on, through the list. On the first try, he had one correct answer, *Iron*. Milner reread the list of word pairs, and then tested him again. The second time he recalled *Cries*, *Iron*, and *Flower*, all belonging to the easy pairs. The third, final round, he remembered *Apple*, *Cries*, and *Iron*. He could not consolidate the difficult pairs. A half hour later, Henry retained the one association that he had correct on each of the three trials: *Metal–Iron*. The other associations had melted away because Henry's brain lacked the medial temporal-lobe infrastructure required to consolidate and store them.[20]

In their groundbreaking 1957 paper detailing Henry's operation and his psychological test results, Scoville and Milner launched the modern era of memory research. Although studies of previous patients, especially F.C. and P.B., had suggested that the hippocampus was critical for the establishment of long-term memory, Henry's case clinched the connection. As a result of his consistently poor performance on numerous and diverse memory tests, the hippocampus became the focus of thousands of memory researchers all over the world.[21]

We now know that consolidation is an essential aspect of memory, but how exactly does it occur? What are the underlying processes in the brain? Answering these questions is essential to understanding Henry's memory impairment.

Memory consolidation depends on dialogues among brain circuits, coupled with cellular changes within networks of cells, specifically those in the hippocampus. It requires intense conversations between the hippocampus and areas in the temporal, parietal, and occipital cortices where bits and pieces of memories are stored. These communications between neurons reorganize and strengthen connections among memory-processing regions to ensure that information is preserved in the cortex.[22]

Messages travel from each neuron to its neighbors by way of a long tail, an *axon*. At the end of the axon, the message, coded in electrical and chemical signals, crosses the gap between the neurons, known as a *synapse*. The synapse contains a *synaptic cleft*, a corridor through which molecules travel from one cell to the next. On the other side of the synapse, *dendrites*, treelike branches of an adjacent neuron, receive messages and pass them securely to the body of that neuron for processing. Each neuron has its own output, the axon, and many inputs, the dendrites (see Fig. 8).

In the middle of the twentieth century, scientists began to hypothesize about the connections between neurons. In 1949, the Canadian psychologist Donald O. Hebb speculated that a structural memory trace in the brain is the foundation for long-term memory formation. Hebb's idea was that learning causes growth in brain structures, thereby establishing memory traces. His thinking was influenced by Spanish anatomist Santiago Ramón y Cajal who in 1894 wrote that "mental exercise" likely resulted in growth on axons and dendrites. Hebb adopted this idea and took it further. In considering what happens at a synapse when one neuron talks to its neighbors, Hebb proposed that when one cell excites another repeatedly, tiny structures on both sides of the synapse swell. (In current terminology, the structures on the axon are called *axonal varicosities* and those on the dendrites are called *dendritic spines*.) This growth makes it more likely that, in the future, the first cell will activate the second cell again. When animals and humans learn new information, several cells in close proximity are repeatedly excited at the same time, forming a closed circuit that strengthens gradually as learning progresses. This postulate, known as *Hebb's rule*, pinpointed the synapse as a critical location to uncover the physiological basis of learning and memory. At that time, Hebb had no direct physiological proof that closed pathways, or loop circuits, participated in behavior or learning. As it turns out, however, his visionary hypothesis about the brain's flexibility was correct—Hebbian plasticity does exist, and Hebb's influence continues as neuroscientists investigate whether this kind of plasticity is, in fact, responsible for learning and memory.[23]

This line of inquiry advanced considerably in the late 1960s with the discovery of the phenomenon of *long-term potentiation (LTP)*, which many neuroscientists now believe to be the physiological underpinning of learning and memory. In 1966, Terje Lømo, a PhD student at the

University of Oslo, performed experiments on anaesthetized rabbits to explore the role of the hippocampus in short-term memory. When Lømo applied a series of fast pulses of electrical stimulation to the axons that carried information into a rabbit's hippocampus, he found that after each successive zapping, neurons on the other side of the synapse in the hippocampus responded to the same input more quickly, more strongly, and in greater number than they had previously. The stimulation strengthened the transmission of information from one cell to another, much like turning up the volume on a radio. A critical characteristic of the enhancement was that it lasted for more than an hour. Lømo called this new finding *frequency potentiation*, and showed that it could be induced by repeatedly activating the axon of one cell, which generated a signal that crossed the synapse and caused a firing increase in the cell receiving this input. After further studies in rats, researchers in the early 1970s changed the name of this phenomenon to *long-lasting potentiation*, and later in the same decade to *long-term potentiation*.[24]

The discovery of LTP provided a versatile model for the study of memory formation in many species of animals. Thousands of researchers on several continents continue to explore the molecular and cellular mechanisms by which specific patterns of activation (that is, different experiences) alter the strength of connections between neurons. LTP provides striking evidence of neuroplasticity, the brain's ability to change with experience. Neuroscientists cite two key concepts in studying the changing brain: *structural plasticity* and *functional plasticity*. Examining structural plasticity has shown us that the anatomy of the hippocampus is not rigidly fixed for life: dendrites and their synapses change continuously in response to experience. Functional plasticity illustrates the property of synapses in the hippocampus and other brain areas to increase or decrease in strength—in essence, the capacity for the activity of one neuron to excite other neurons. At the core of memory is the brain's ability to change as the result of experience, and LTP is an excellent laboratory example of both structural and functional plasticity.[25]

A wealth of research conducted after the discovery of LTP fleshed out its three basic features. First, potentiation is lasting and may persist from a few hours to a few days, and even up to a year (*persistence*). Second, potentiation is restricted to neural pathways that are active when the delivery of specific patterns of stimulation begins the process of encoding new information (*input specificity*). Third, the neuron on the sending side of the synapse and that on the receiving side of the synapse must be active simultaneously (*associativity*).[26]

In the mid-1980s, a major question remained unanswered: is LTP responsible for learning, as observed in the performance of an animal engaged in a memory task? In other words, could learning and memory take place when researchers *prevented* LTP from happening? A 1986 study addressed this question and provided further evidence that deficient LTP is linked to spatial amnesia. Neuroscientists at the University of Edinburgh, in collaboration with colleagues at the University of California, Irvine, trained normal rats to swim to a platform hidden in a pool of opaque water (later christened the Morris water maze). After a few days of practice, one group of rats figured out where the platform was located in the pool and was able to use it to climb out. In another group of rats, the researchers delivered a drug that shut down LTP in their hippocampi as they tried to learn the task. These rats had trouble finding the platform. This result, a clear indication of a deficit in spatial memory related to the blockade of LTP, is similar to Henry's trouble finding his way back to his new house after his family had moved.[27]

A major advance over the pharmacologic blockade method came in 1996 when several papers from the laboratories of two Nobel Laureates, Susumu Tonegawa and Eric Kandel, heralded a revolution in the quest to understand the kind of learning and memory that depends on the hippocampus. These distinguished researchers and their numerous collaborators used the powerful *gene knockout* technology to exclusively remove the NMDA (N-methyl-D-aspartate) gene from a specific kind of neuron—*pyramidal cells*—located in three separate parts of the mouse hippocampus. When LTP occurs, active cells on the sending side of synapses release a neurotransmitter, glutamate, which opens doors called NMDA receptors on the receiving side of the synapse—*if* these receiving cells are simultaneously active. This communication initiates processes that result in the protein synthesis and structural changes that help nail down the to-be-remembered event and make the synapse more effective (potentiated).[28]

By disabling the NMDA gene selectively in one area at a time, Tonegawa, Kandel, and their colleagues could describe what roles this gene played in memory formation, and they found that the CA1 and CA3 areas in the hippocampus had distinct specializations. When the NMDA receptor deletion targeted CA1, the mice were impaired on the Morris water maze, requiring more time than sibling mice with intact NMDA receptors to swim to and climb up on the hidden platform. By contrast, the researchers discovered that another part of the hippocampus, the CA3 region, plays a different role in memory. There, NMDA receptors were necessary for *pattern completion*, which occurs when animals have to retrieve an entire memory after they are given only a fragment of that memory as a cue. These experiments using the genetic blockade method signified a major advance because they targeted specific cells in the hippocampus and conclusively established the role of receptor-dependent synaptic plasticity in spatial memory.

But do humans exhibit LTP? Since the late 1990s, labs in Germany, Austria, Canada, Australia, and England have been inducing LTP in the human hippocampus, motor cortex, and spinal cord. In some people, LTP may be maladaptive, either when it is enhanced or decreased, and it is possible that some neurological and psychiatric disorders result from too much or too little LTP. This possibility opens up numerous treatment options for the millions of people made miserable by these conditions. An outstanding characteristic of the human brain is its plasticity—its capacity to change with experience. By capitalizing on this potential, it should be possible to correct dysfunctional LTP. An exciting prospect is that memory loss, epilepsy, chronic pain, anxiety, addiction, and other conditions that have been linked to dysfunctional LTP can be reduced by deploying, at strategic points in the nervous system, one of the many chemicals that control LTP.[29]

It appears more and more plausible that LTP is necessary for learning, but there is much we do not understand. So far, scientists have been unable to prove that LTP is as enduring as our memories; LTP lasts for weeks at most, whereas long-term memories may last for decades. Neuroscientists are also working to understand how the cellular-molecular mechanisms that they have observed directly in the laboratory relate to the encoding, storage, and retrieval of specific memories in everyday life. We have a lot of ground to cover before we can bridge the gap between processes in our hippocampal cells and how well we do on a written test for a driver's license.

Some people deny they dream, and others say they always forget their dreams. This is because fully recollecting our dreams requires some work. To document our dream content, we have to have a pad and pen next to our bed so we can record each dream immediately upon awakening, before it slips away. Dreams typically incorporate experiences from our past and may also play a role in memory consolidation. Still, we do not yet have direct evidence that dreams are *required* for memory consolidation, so we must interpret experiments relating dreams and memory cautiously.

To better understand how memories are consolidated—anchored—researchers in the mid 1990s embarked on sleep studies in rats. These experiments documented mental content during sleep by means of electrodes placed in the rats' hippocampus. The specific patterns of neural activity recorded during sleep were compared with recordings made when the same animals were awake. This comparison often showed a clear correspondence between the two sets of recordings, giving some insight into the role of sleep in remembering awake experiences.

This research evolved from the major discovery of *place cells* in the hippocampus. In 1971, neuroscientists at University College London identified specialized neurons in the rat's hippocampus that signal the animal's current location in space. Each place cell corresponds to a particular zone in the rat's space—the cell's *place field*. When the cell fires, it signals to the rat where it is and the direction in which it is facing. Place cells are activated, for instance, when a rat is placed in a maze and has to find the way to get a treat. Together, these cells map the rat's environment. Place fields provide the best example we have of how the rat's world is represented inside its hippocampus.[30]

Since this discovery, place cells in rats and mice have attracted enormous experimental interest. When these animals perform in maze experiments, the cells fire in patterns and sequences that correspond to different locations in the maze, pinpointing where the animal is running or stopping. Even more intriguing is that these place cells are reactivated in the same order *after* the rat is removed from the maze. That is, when the rats are quiet—sleeping or pausing—their place cells replay the pattern of neural activity that occurred during a previous trip.

How does the activity of place cells affect the formation of long-term memories? In 1997, neuroscientists at the University of Arizona proposed that the hippocampus facilitates the reactivation of cortical activity patterns during offline periods, such as sleep or quiet wakefulness, when the cortex is less engaged in processing incoming information. To explore this possibility, the researchers inserted electrodes next to place cells in the rats' hippocampus to record their activity. Each recording session had three phases—sleep, maze running, and sleep. The researchers predicted that during the second sleep period, neuronal firing in the hippocampus would resemble that in the cortex, and, in addition, the pattern of activity would correspond to that during maze running. The results verified their predictions. During sleep, the patterns of neuronal firing when the rat was in the maze were re-expressed in the hippocampus and cortex, and the mental representations, the paths, in the two areas resembled the maze-running representation that preceded the second sleep period. This correspondence suggests that circuits in the hippocampus and cortex were interacting during sleep, but the question lingered: Did this transient activity play a role in long-term memory?[31]

With the Arizona study as a springboard, MIT neuroscientist Matthew Wilson and his colleagues conducted experiments in freely moving rats and mice, from which they simultaneously recorded the firing of about one hundred cells. The main question driving this research was, how do large populations of cells in the hippocampus form and retain memories? The answer came from recordings of the activity of place cells in the animal's hippocampus. These animals wore little hats that held a number of miniature recording electrodes—*tetrodes*—that allowed the scientists to eavesdrop on many neurons simultaneously. Together, these data gave researchers a realistic picture of the electrical activity of numerous place cells at any given moment.

To compare brain activity during sleep and wakefulness, researchers monitor EEG recordings and divide sleep into separate stages characterized by different electrical activity. In humans, *slow-wave sleep*, a deep sleep, is most common during the first half of the night, and *REM* (*rapid eye movement*) *sleep*, a lighter sleep, occurs mainly during the second half. Animals also experience slow-wave and REM sleep.[32]

In a 2001 experiment, the Wilson Lab recorded the activity of place cells in the rats' hippocampus, first for ten to fifteen minutes as the animals ran a maze to get a treat, and later from the same cells for one to two hours as the rats were sleeping. When the researchers compared the behavior of cells in the hippocampus during maze running and during subsequent *REM* sleep, they found a remarkable correspondence—the two sets of data were strikingly similar, suggesting that the sleeping rats were replaying the behavior learned previously in the maze. Hippocampal neurons fired in the same order during REM sleep as they had during learning.[33]

These patterns of neural activity, representing a behavioral sequence, lasted as long as the real experience did. Repetition of the firing patterns of a collection of individual cells is known as *memory replay*. The awake rats encoded the sequences in a specific part of their hippocampus (area CA1), and this electrical activity was still detectable twenty-four hours later during REM sleep. This recapitulation of neural activity linked to a previous behavioral experience is strong evidence of a persistent memory. CA1 place fields encoded the location of the animal in space, and from this information, its brain assembled sequences of locations. Cross talk between the hippocampus and brain areas in the cortex that are specialized for spatial abilities likely contributed to this achievement. The replay of awake activity during REM sleep may enhance memory consolidation in the cortex as a result of interactions between cortical and hippocampal circuits. This and many other hypotheses still need to be explored before we can definitively nail down the role of memory replay in learning and consolidation, but a large body of evidence supports the view that hippocampal place cells remember.[34]

Further evidence that memory replay is an important component of learning and long-term memory came from Wilson and colleagues' examination of brain activity during slow-wave sleep. They found that the effect on memory was different from REM sleep. They trained rats to run to and fro on a track, rewarding them with a piece of chocolate every time they reached one end or the other. As the rats indulged their sweet tooth, the researchers recorded the activity of many cells in the hippocampus simultaneously. They then monitored the same cells while the animals slept. The rats' awake experience was replayed in the hippocampus during slow-wave sleep, only now at high speed—a four-second lap on the track was replayed fifteen to twenty times faster in the brain during slow-wave sleep. The memories that were reactivated reflected the order of the events, not their actual duration. These ordered events could then be played back over a shorter time than the animal needed to execute them, making them appear compressed.[35]

If we create a picture in our mind's eye of the route we take from our home to the supermarket, and then imagine ourselves traversing this route, our mental trip will take much less time than it actually takes to go by car. We no doubt compress our dream content in the same way. In Wilson's experiment, this kind of memory replay was most likely to occur in the first few hours of slow-wave sleep following the experience, so its role may be in the early processing of information, prior to the establishment of long-term memory.[36]

Because memories are stored throughout the cortex, cortical activity must be part of the process of memory consolidation that occurs during sleep. We know that the hippocampus and cortex must work in partnership to receive, organize, and retrieve knowledge, but how do remote brain areas collaborate? Wilson's team contributed another important insight in 2007 when they observed a close association, during memory replay, between cellular activity in the hippocampus and in the cortex.

The researchers trained normal rats to run in a figure-eight-shaped maze. The rats began each experimental session in the middle of the eight and, to receive a food reward, had to run alternately to the left and to the right. After three weeks of training, the researchers implanted small electrodes in the rats' hippocampus and visual cortex—a sensory processing area in the back of the brain that receives information from the eyes—and recorded patterns of neuronal firing from the two areas. The scientists found evidence of memory replay during slow-wave sleep in both, suggesting that rats, like people, have visual dreams. The activity patterns in the hippocampus resembled those in the visual cortex.[37]

It would be wonderful if all we needed to have better memory was a good night's sleep packed with memory replay. Although that possibility is remote, growing evidence shows that memory consolidation and synaptic plasticity benefit from sleep. Experiments in humans that examine the effects of different sleep stages on consolidation now focus on the links between particular kinds of memory performance—declarative and nondeclarative—and the type and duration of sleep. This knowledge enhances our understanding of memory, allowing us to go beyond behavioral experiments by looking inside the brain for the neural processes that support consolidation. By examining the many physiological changes that accompany sleep and sleep disorders, researchers may be able to formulate novel remedies for insomnia and memory impairment.

Henry's deficit was in consolidating, storing, and later retrieving new facts and events. Still, his amnesia could not necessarily be blamed on a deficit in *retrieval*: he could still recall facts he had consolidated and stored before his operation. He loved to talk to our team at the CRC about his preoperative experiences and family life. When he retrieved these memories, however, he could not integrate the old memory traces with information in his current life. For instance, when he talked about his gun collection, he could not update this narrative by saying what had happened to it. Henry had consolidated information before his operation but could not reconsolidate it after, so his childhood memory traces remained engraved in his brain, unrefreshed.

Think of *reconsolidation* as a memory-updating process. If you unpack a suitcase and then repack it, the clothes will be arranged slightly differently than they were before, and you may leave out some items and add others. Retrieval and reconsolidation of old memories suddenly makes them labile, a state in which they are again vulnerable to distortion and interference. At this time, memories can be modified by new information.

If I ask you when you last had Chinese food, you initiate a mental search based on internal thoughts of food, dinner, leftovers, chopsticks, and so forth. You might be driving down a street in Chinatown, and that environment could help revive your memories too. When you recall the meal I asked you about, the process of retrieval reactivates consolidation mechanisms similar to those that occurred at the time of the original meal, even if it was years ago. The memory will be altered by your current thoughts. For instance, finding out a year after you ate the Chinese meal that you were allergic to MSG modified the memory of the original experience by incorporating the cause of that night's splitting headache. Also, having a similar meal, such as celebrating your last birthday at a Korean restaurant, could interfere with your memory. Anything you think about at the same time as you mentally revisit the Chinese meal helps shape the new memory. But while you may expose your memory of the event to distortion each time you retrieve it, you also make the edited version more likely to stick in your mind. Reactivating an old, previously consolidated memory creates new memory traces, and this mushrooming of traces makes older memories more resistant to interference from other brain activity. If you reactivate the memory of that meal every month for the next six months, the new memory will be more robust and more likely to survive over time, although its resemblance to the initial episode may diminish.

Retrieval is a reconstructive process—a more complex operation than simply activating the appropriate memory traces. Memories change every time you retrieve them. No two memories are alike because the process of calling them up and engaging them changes the content of the memory that you put back in storage. Each time you recall your last birthday, the details are slightly different—some are deleted and others are added. Consolidation occurs anew. The realization among neuroscientists that consolidation happens again, during and after retrieval, has been named the *reconsolidation hypothesis*. The basic idea is that memories are taken out of storage—retrieved—and then put back again—reconsolidated. Remembering is a mixture of the information previously stored in long-term memory with the information in an immediate situation.[38]

Whenever a memory—good or bad—is retrieved, the new information in the retrieval environment must be integrated into the existing, well-established network underlying that particular memory. When a memory is reconsolidated, new content is interleaved with existing content, leaving the reconsolidated memory embellished and changed. *False memories* are a striking example. Neuroscientists at the University of California, San Diego, questioned college students as to how they heard about the O. J. Simpson trial verdict, whether from radio, television, or a friend, three days after the announcement. Some of the students were tested again fifteen months later and gave relatively accurate answers; others were retested thirty-two months after and were less accurate despite being more confident in their answers. At both time periods, the students misremembered how they heard the information, supporting the view that consolidated memories are unstable and can be edited.[39]

In 1997, two neuroscientists at l'Université Pierre et Marie Curie in Paris found physiological evidence to show that memory can be changed. They trained rats on a maze that had eight arms radiating from a small central platform. At the end of three arms, they placed Cocoa Puffs cereal to entice the rats to choose those three arms and not the five unbaited ones. Once a day, the researchers placed each rat in the center of the maze and allowed it to visit the alleys as it wished. The trial ended after it visited the three alleys with the treats. After a few days, the rats learned this trick perfectly, entering the three alleys with the Cocoa Puffs immediately and not bothering with the five empty ones.[40]

To test the strength of the rats' newly consolidated memories, the researchers reactivated the memory by allowing the animals to perform a single errorless trial. They then immediately injected the rats with

Dizocilpine, a memory-degrading drug that blocked the activity essential for consolidating their experience on the single errorless trial. Twenty-four hours later, the rats did not remember which arms contained the Cocoa Puffs.

The researchers demonstrated reconsolidation using a clever experimental strategy. They gave the rats one additional test session, and that is all the practice they needed to regain their pre-drug proficiency in the maze. Their memory for the maze had been disrupted but not totally lost. The extra test session gave the rats information that they reconsolidated with the established memory, enabling them to perform once again at their pre-drug level.[41]

This experiment established that when a memory is reconsolidated, at least some of the cellular events that occurred during the initial consolidation are reenacted. The evidence that long-term memories are changed each time they are retrieved reinforces the view that memory is an ongoing dynamic process driven by life's events. Further biological confirmation supports the hypothesis that recapitulation of a long-term memory makes it stronger and more stable.

Reconsolidation occurs constantly in your everyday life. Say that, after an absence of ten years, you visit the community where you grew up. It will strike you that what you see does not exactly match what you remember; the neighborhood may look, sound, and even smell different. As you retrieve long-term memory traces of your old stomping grounds, you update them by incorporating new features of the landscape and evaluating your roots from an adult perspective. In doing so, you are taking advantage of the lability of the old traces to establish more current and stronger memory traces. This updating process happens because a mismatch occurred between your old consolidated memory and the new information. You can adjust your opinion of people in the same way: a colleague who made a bad first impression may turn out to be a respected and valued partner.

The concept of reconsolidation was first proposed in the late 1990s, and it has the potential to bring about important breakthroughs, for instance in the treatment of post-traumatic stress disorder (PTSD). Many

Iraq War veterans have been crippled by PTSD to the extent that they cannot resume their prewar lives; they persistently re-experience traumatic events, with symptoms including sleeplessness, irritability, anger, poor concentration, and constant anticipation of danger. A team of researchers has been testing a method to reduce the pain of PTSD without wiping out the memory of the distressing incident altogether. Their approach is to give patients Propranolol, a drug that dampens the activity of the sympathetic nervous system, the body's tool for physically expressing emotion. This drug selectively blocks the reconsolidation process for the emotional content of the event, but not for the facts themselves. After even a brief treatment with Propranolol, patients may feel better; they can still recall the details of the traumatic events, but without extreme mental anguish. This work suggests that during retrieval and reconsolidation of a traumatic memory, Propranolol reduces activity in emotion-related areas, but does not interfere with function in the hippocampus where the basic facts are reconsolidated.[42]

In our daily lives, memory for unique events, episodes, also benefits from reconsolidation. Reactivating one element of a complex episodic memory also reactivates many other memories linked to that event. For example, I remember that on my first day at MIT, the administrative officer in our department showed me how to use the Xerox machine. He told me to put my hand down on the glass and pressed a button, and after a few seconds, out came a piece of paper displaying an image of my hand. I was impressed and walked away from the marvelous machine with my first Xerox copy. When I called up my Xerox machine memory from 1964, I unleashed a barrage of recollections related to that day at MIT. If you frequently retrieve the details of a particular event in your life, like your first day on the job, then your memory of this event will be more reliable than if you had allowed time to pass without retrieving the event at all.

Studying how Henry forgot gave us a better understanding of how we remember. More than a hundred years ago, Müller and Pilzecker first proposed that memories consolidate over time and that partially consolidated memory traces are vulnerable. In 2004, a psychologist at the University of California, San Diego, took this observation further and proposed a coherent theory of *forgetting* based on converging evidence from three disciplines—psychology, psychopharmacology, and neuroscience. Forgetting occurs because we are constantly forming new memories that interfere with other memories still undergoing consolidation. During this probationary period, memories can be degraded by mental exertion of any sort, whether it is related or unrelated to the incompletely consolidated memories.[43]

Imagine a forty-five-year-old professional embarking on a one-week vacation at a tennis resort. On the long drive from his home, he thinks back to the meeting he had with his supervisor the afternoon before, in which she asked him to lead a new project in his area of expertise. Recalling her specifications, he starts to plan out his approach to the new project, making a mental outline of his proposal and then, as the miles fly by, pondering ways to implement each stage of the venture. At last, he arrives at the resort, and his mental work time comes to a halt. He is greeted by his hosts and immediately gets caught up in the fun. He sees unfamiliar faces, hears new names, and tries to absorb instructions about what he will do when, where, and with whom. He happily wheels his suitcase and tennis bag to his room, unpacks quickly, and heads for the pool. He settles into a different world and lets his everyday life fade. For the next seven days, he focuses on his forehand, two-handed backhand, serve, overheads, and net game. All the time, unbeknownst to him, his hippocampus has been actively recording the novel sights, sounds, conversations, and places, and creating a mental map of the resort. A week later, when he gets back in his car to drive home, he likely will have forgotten many details of his meeting with his supervisor and his brilliant plans for the new undertaking. The barrage of information from his vacation interrupted consolidation of the information he had encoded the week before. Newly acquired memories are easily altered, and interference from subsequent events can erase them partially or completely.

Forgetting with the passage of time is normal; remembering an event that happened ten years ago is more difficult than remembering one that happened last week. We forget the past because newer activities and thoughts push those old memories aside. Loss of memories over time may also occur when the initial episode was poorly encoded. Having Henry as a willing research participant gave my lab a unique opportunity to address these topics and determine whether damage to structures in the medial temporal lobe exacerbated forgetting.

In 1986, we designed a series of experiments with Henry, then fifty-eight years old, to clarify the circumstances under which forgetting occurs. Conventional wisdom held that amnesic patients forgot information more quickly than healthy individuals. To test this theory, we showed Henry and a control group 120 slides, each with a different complex, colorful magazine picture, depicting animals, buildings, interiors, people, nature, and single objects. Henry viewed each picture for twenty seconds and, as instructed, tried to remember them. In subsequent testing, we asked him to look at two pictures side by side, one studied and one new, and to say which one he thought he had seen before. We call this kind of memory retrieval *recognition memory*—consciously choosing between two possible answers, one of which is right.[44]

We tested Henry's memory for the 120 pictures in four sessions. The tests occurred at different intervals after his initial exposure to all 120 pictures. Henry saw thirty pictures after ten minutes, a different thirty after one day, a different thirty after three days, and the final thirty after one week. For us to compare Henry's rate of forgetting with that of the control participants, his scores after ten minutes had to be identical to theirs. We achieved this critical parity by allowing Henry, in the initial learning phase, to see the pictures for twenty seconds, whereas control participants had only one second to encode them, thus compensating Henry with nineteen extra seconds to encode each picture.

To everyone's surprise, the information that Henry encoded in twenty seconds allowed him to recognize the pictures as well as, if not better than, the controls after one day, three days, and one week. More astonishing still was the discovery that Henry's recognition memory was normal six months after his initial exposure to the pictures. Crucially, when compared with healthy adults of his age and education level, he did not show faster forgetting of complex, colorful pictures.[45]

How could Henry achieve normal picture-recognition scores when most other indices of his long-term memory pointed to failure? He remembered virtually nothing about his everyday life and bottomed out on every other test of new declarative memory. I was at a loss to explain the picture recognition result theoretically. Describing the experiment to colleagues, I speculated that Henry just had a vague feeling in the pit of his stomach that helped him decide in one of two ways. On each trial, he had to choose between two pictures—one that was in the study list and a new one. Thus, he based his response either on accepting one of the pictures as familiar or on rejecting one of the pictures as familiar and choosing the other one. This process occurred automatically and was based on the strength of the memory for one picture compared to another. We could not ask Henry to describe what he was thinking because constant interruptions would have compromised the experiment. But it is likely that Henry was automatically engaging cortical areas in the back of his brain that are known to have a massive storage capacity for visual material.

Around the time we were performing these tests, researchers in mathematical psychology capitalized on information-processing theory to shape a solid theoretical framework for recognition memory. This framework was rooted in a model that Richard Atkinson and James Juola proposed in 1974. Their long-term memory tasks required participants to memorize as many as sixty words. In the test that followed, participants saw one word at a time and had to decide whether the word was one they had memorized. If the participants answered quickly, their response, based only on a feeling of familiarity, they had a high probability of being wrong. If they took additional time, however, they had the opportunity to explicitly recall that the word was in the study list, with a high probability of their being correct. Based on their results, the researchers described the process of recognition in normal individuals as consisting of two independent retrieval processes. These ideas about recognition memory were formalized in 1980 when cognitive psychologist George

Mandler introduced the now-popular dual-process model of recognition. He formalized two kinds of recognition memory—*familiarity* and *recollection*. Subsequent studies by another cognitive psychologist elucidated a fundamental distinction between these two uses of memory, consistent with the 1974 model—familiarity relies on fast, automatic processes, whereas recollection is a slower, intentional use of memory that demands a person's attention.[46]

We have all had the experience of seeing someone on the street whose name we do not know, but with a vague sense that we have met him before. This kind of recognition is based on *familiarity*. It is not attention demanding, and it happens automatically. Henry engaged this process when he identified the complex magazine pictures he had seen, a task with a relatively small cognitive demand. Conversely, when we meet an old friend on the street, we easily recall in great detail the good times we have shared. This kind of recognition is based on *recollection*, a process that demands effort and attention for searching our memory stores. Because this process depends on intact hippocampi, Henry was unable to engage it in his daily life and on most formal tests of his memory.

Henry's unexpected test results in the 1980s revealed a crucial fact about the division of labor in the brain—familiarity and recollection are managed by independent processes in separate brain circuits, one preserved in Henry's brain and the other destroyed. This observation was later clarified by behavioral evidence from hundreds of sources linking recollection with processes in the hippocampus and familiarity with processes in the perirhinal cortex. Although the hippocampus and perirhinal cortex are close, interconnected neighbors, they each make unique contributions to memory retrieval.[47]

This distinction helps us understand Henry's ability to recognize complex pictures up to six months after he encoded them. His operation removed part but not all of his perirhinal cortex, so Henry possibly recognized the pictures in our experiments by engaging his partially spared perirhinal cortex in collaboration with other normal areas in his cortex. Still, this partnership was not sufficient to support his long-term memory in everyday life. Our picture-recognition finding was a remarkable exception, in stark contrast to his performance on other declarative memory tests where his recognition memory was consistently deficient.[48]

Subsequent functional MRI studies supported our hunch about Henry's sense of familiarity for the complex pictures we had shown him. In 2003, a team of cognitive neuroscientists in California showed that familiarity and recollection depend on different anatomical areas within the medial temporal lobes. Inside the MRI scanner, their research participants encoded words such as *NICKEL* displayed in red letters and other words such as *DEER* displayed in green letters. After the scan, participants took a recognition memory test in which they viewed a random mix of studied words and unstudied words, and gave two responses. They first rated how confident they were that they had seen, or not seen, each word before. They then decided whether the letters in each word had been red or green when they first saw them in the scanner, a measure of their *source memory*. The ability to remember the color of the letters, source accuracy, assessed recollection—consciously associating each word with its color. Source memory judgments could not be based on familiarity because the study list contained red words intermixed with green words, making them equally familiar or unfamiliar at the time of the test.[49]

The researchers analyzed each participant's functional MRI images individually. They distinguished brain circuits that showed increased activation when participants encoded words they later recognized based on *recollection*, versus brain circuits that showed increased activation when participants encoded words they later recognized based on *familiarity*. The contribution of familiarity increased gradually as recognition confidence increased. The more confident the participants were that they had seen a particular word before, the greater the familiarity effect.[50]

Consistent with the theory that the perirhinal cortex and hippocampus play different roles in recognition memory, the functional MRI analyses uncovered two distinct circuits, one for each kind of recognition memory. The researchers found a circuit dedicated to the feeling of familiarity in two contiguous areas, the entorhinal and perirhinal cortex. Here, brain activity increased as familiarity increased. Two other areas showed heightened activity when participants correctly remembered the color of the letters, indicating accurate memory for source information—the color—an index of recollection memory. This hotspot was in the back of the hippocampus and in the cortex next to it, the parahippocampal cortex.[51]

These findings indicate that the hippocampus and parahippocampal cortex specialize in recollection, whereas the perirhinal and entorhinal cortices specialize in familiarity. Henry's anatomical MRI scans showed that he had some perirhinal tissue remaining on both sides of his brain. We reasoned that his residual perirhinal areas jumped into action when we asked him to remember the complex magazine pictures, enabling him later on to select the ones he had seen before, based on whether they seemed familiar.

Henry's case proved that hippocampal lesions cause profound difficulty with recollection, and a parallel question arose with respect to the perirhinal cortex and familiarity. Would another person with a lesion restricted to the perirhinal cortex exhibit a deficit in familiarity? The answer came in 2007 from a patient who showed impaired familiarity and preserved recollection after she sustained damage to her perirhinal cortex but not her hippocampus. A group of Canadian researchers examined recognition memory in this patient, N.B., who had undergone a left anterior temporal lobectomy to relieve intractable epilepsy. Her operation was atypical because, unlike F.C., P.B., and Henry's, it spared her hippocampus while excising a large portion of her perirhinal cortex. N.B.'s performance on recognition memory tests was the opposite of Henry's—recollection was normal, while familiarity was impaired. This striking case report adds weight to the theory that separate circuits in the medial temporal-lobe region support recollection and familiarity. Still, researchers continue to debate the precise localization of recollection and familiarity processes and have deemed it a topic worthy of further study.[52]

Henry's declarative memory was broken, and he was left with only vague feelings of familiarity. He could not evaluate whether these mental impressions were trustworthy, but perhaps that did not matter because they provided content in his life. Henry's preserved capacity for familiarity helped him during his twenty-eight years at Bickford. He felt comfortable in the homey atmosphere, and one staff member described him as "the mainstay of the lounge." He was very popular with the other patients, and some of them asked for him by name. His kind heart and polite demeanor helped him to be tolerant of the demented people who surrounded him. As his cordial interactions with them clearly showed, Henry certainly did not regard his fellow patients and the Bickford staff members as strangers.

One advantage I had in my interactions with Henry was that my face was familiar to him. He believed we went to high school together, so I was not an outsider. He made the same association with a few of the female staff members at Bickford with whom he regularly interacted, and this repeated exposure over time strengthened the feeling that he knew some of these people. Even though the faces, objects, and technology in his environment changed dramatically decade by decade, Henry accepted these changes without question, incorporating them into his universe. Before his operation, he had watched television programs in black and white; after his operation, when color television became available, he did not comment on the dramatic difference. Likewise, in our lab, he was so comfortable sitting in front of computers to perform tests that it seemed like they had always been part of his life. The sense of familiarity that permeated Henry's world helped him cope with his disabling amnesia by grounding him and giving him the feeling that he was among family at Bickford and MIT.

Eight

Memory without Remembering I

Motor-skill Learning

Henry's brain damage was restricted to his medial temporal-lobe structures, and the remaining areas, except for his cerebellum, still operated normally. These other regions supported several kinds of unconscious learning. In everyday life, he could acquire new skills and remember how to perform them.

One skill Henry needed to learn as an older man was how to use a walking frame, which he came to depend on due to the side effects of his antiseizure medication. Although his operation had the desired result of markedly reducing the number of grand mal seizures he experienced, Henry still had to take epilepsy drugs. He had been taking high doses of Dilantin before his operation and continued to take therapeutic doses until 1984, when a neurologist recommended that he switch to a different antiseizure drug. By that time, Dilantin had caused several damaging side effects, including osteoporosis, which led to several bone fractures. Dilantin also resulted in significant withering of his cerebellum, the large structure at the back of the brain responsible for maintaining balance and coordination. As a result of this brain shrinkage, Henry was unsteady on his feet and moved slowly. Another of his seizure medications, Phenobarbital, is a sedative and likely contributed to his overall slowness.

Henry's osteoporosis progressed to the point that it was unsafe for him to walk on his own. In 1985, he fractured his right ankle, and in 1986, had his left hip replaced. During his recovery, his doctor prescribed a walking frame to keep him physically active and safe when on his feet. Once he received this new tool, he had to learn several new procedures to use it properly. With practice, Henry acquired the technique for walking, transferring his body from a chair to his walker, and returning to a chair. When I asked him why he used the walker, he replied, "So I won't fall down." He had no conscious, declarative knowledge that he developed osteoporosis as a result of taking Dilantin; nor did he remember that he'd had several fractures that had required hospitalization and rehabilitation. But Henry did retain the new motor skills from day to day and month to month, a striking example of his ability to obtain and hang onto procedural knowledge.

In the lab, formal demonstrations of his motor-learning ability echoed these everyday achievements. Henry recruited areas in his brain that were spared, and he could learn and remember without knowing that he was doing so. The use of the term *memory* in this situation underscores the point that we possess more than one kind of memory—we engage our conscious, *declarative* memory processes when we recall what we need to buy at the grocery store, whereas we rely on our unconscious, *nondeclarative* memory when we can still ride a bicycle after not having done so for ten years.

Recognizing that learning can take place without awareness was one of the most significant advances in human memory research. In the twentieth century, much of the scientific research on amnesia focused on declarative learning and memory, but a parallel story unfolded, revealing a different kind of memory, nondeclarative learning, by which amnesic patients could perform new tasks despite the inability to explicitly describe their learning experience. Nondeclarative learning is sometimes referred to as *procedural* or *implicit*. A broad range of preserved learning capacities is covered under the nondeclarative umbrella—motor-skill learning, classical conditioning, perceptual learning, and repetition priming. These procedures differ in several ways, including the number of trials required for acquisition, the critical brain substrate, and the durability of the knowledge.[1]

The first account suggesting that learning could occur in an amnesic patient appeared in 1911. Édouard Claparède, a psychologist at the University of Geneva, related a remarkable clinical anecdote about a forty-seven-year-old woman whose memory was impaired due to Korsakoff syndrome, amnesia attributed to thiamine deficiency. Like Henry, she retained the general knowledge of the world she had acquired before the onset of her illness; for example, she could name all the European capital cities and do simple arithmetic in her head. She could not, however, remember a list of words or stories read to her and did not recognize the doctors who cared for her.

To explore her capacity for learning, Claparède shook his patient's hand one day with a pin hidden in his palm. She felt the pinprick and recoiled. When he approached her the next day with his hand outstretched, she declined to shake his hand but had no idea why. Clearly, she took in information at the time of the handshake, but the next day she could not bring to mind her unconscious memory of the painful experience that guided her response. She could not state her fear, demonstrating that her declarative memory was impaired. But at the same time, she witheld her handshake, indicating that her nondeclarative memory was still functioning.[2]

Four decades later, Brenda Milner provided the first formal experimental demonstration of preserved learning in amnesia. In 1955, when she first evaluated Henry at Scoville's office in Hartford, she tried to unearth any evidence of new learning using many different kinds of behavioral tasks. Her tests were not driven by any particular hypothesis, but the payoff was tremendous: on one of the tasks, Henry's performance improved measurably during three days of practice. This exciting, serendipitous finding suggested that the structures removed from Henry's medial temporal lobes were not necessary for this kind of learning. Milner's experiment suggested that the brain houses two different kinds of long-term memory, one on which Henry failed and another on which he succeeded. The ensuing decades witnessed the publication of thousands of investigations of nondeclarative memory inspired by Milner's discovery (see Fig. 9).

Among the tests Milner chose was a motor-skill learning task, mirror tracing, which she administered on three consecutive days during one visit. Each day, Milner asked Henry to trace a five-pointed star, keeping his pencil inside its borders. This task was challenging because the star, printed on paper, was mounted on a horizontal wooden board hidden from Henry's view by a near-vertical metal barrier that blocked direct vision of the star, his hand, and the pencil. He reached around the right side of the metal barrier and could see the star, his right hand, and the pencil in a mirror mounted on the far side of the wooden board. The entire image was reversed so that if he wanted the pencil to trace around the star away from his body, he had to move the pencil down toward his body. The normal visual cues we use to guide our movements were turned upside-down. The task required mastering a new motor skill—allowing this reversed visual image to dictate the movement of his hand. Every time Henry strayed outside the lines and had to return, it was counted as an error. Most people find the task difficult and frustrating at first but improve over time, and with practice, they gradually trace around the star faster and with fewer errors.[3]

As Henry performed the task over and over again, something remarkable happened. On the first day, his errors dropped steadily from trial to trial, and—unpredictably—he retained what he learned overnight. On the second day, his initial error scores were almost as good as they had been at the end of training the first day, and he continued to trace around the star with fewer and fewer mistakes. On the third day, he performed nearly perfectly—he traced around the star cleanly and rarely veered outside the lines.

Henry had learned a new skill. This learning, however, had taken place outside his conscious knowledge. On days two and three, he had no memory of having done the task before. Milner vividly remembers the last day of testing: after skillfully tracing the star in the mirror, Henry sat up straight and proudly observed, "Well, this is strange. I thought that that would be difficult, but it seems as though I've done it quite well."

Milner speculated that motor skills, such as the one Henry had mastered, might be learned by recruiting a different memory circuit, one outside the hippocampal structures Henry lacked. This unforeseen discovery unlocked a treasure trove of learning processes that do not depend on the medial temporal-lobe circuits that were damaged during Henry's operation, but instead are mediated by brain areas that were left behind.[4]

In 1962, I expanded on Milner's amazing discovery while working in her lab at the Montreal Neurological Institute as a McGill University graduate student. Henry and his mother were in Montreal for a week of testing. By that time, scientists had vetted and verified his declarative memory impairment with tests that required him to remember information presented through vision and hearing. No one, however, had tested to see whether his memory deficit extended to his sense of touch, his somatosensory system. Taking on this project, I presented Henry with a task to learn the correct sequence of turns in a touch-guided maze that he traced with a pen. In Chapter 5, I described his failure to learn the correct route from the start to the finish. But even though his error scores did not decrease over the eighty trials, Henry *had* learned something new. In addition to recording how many errors he made on each trial, I noted how many seconds elapsed between his leaving the start point and reaching the finish. After he and his mother returned to East Hartford, I plotted these data on a graph, and found to my surprise that while his number of errors never changed, his time scores decreased steadily over the same eighty trials. From day to day, he moved more quickly through the alleys in the maze, even though he could not remember the route. This decrease in the time it took Henry to traverse the maze showed that he learned something—the procedure, the *how* to do it. He did not remember the route, but he became increasingly comfortable with the task. This experiment strengthened the view Milner had proposed: motor learning depends on a different memory circuit from the medial temporal-lobe area, which underlies the consolidation and storage of facts and experiences.

Henry's contrasting error and time scores when he traced my tactual maze reinforced the view that free recall—*declarative memory*—depends on the hippocampal region, which he now lacked, whereas skill learning—*procedural memory*—engages different networks that were undamaged. To the best of my knowledge, this result from 1962 was the first quantitative demonstration, within a single experiment, of impaired declarative learning (failure to learn the correct route) with preserved procedural—nondeclarative—learning (improving the motor skill). Further research in patients and in healthy individuals went on to characterize the important distinctions between these two kinds of long-term memory.[5]

To learn a new motor skill, the task must be performed over and over again. Once attained, motor skills are enduring—hence the adage about never forgetting how to ride a bicycle. But as any tennis player can tell you, motor skills are not perfected in a single practice session; instead, they evolve with experience, and performance progresses from the staccato execution of several actions to the integration of those actions into a smooth movement performed automatically. Consider, for example, the many steps required to execute the two-handed backhand. Begin by facing the net with your toes pointed forward and your racket and body in the ready position. When the ball approaches your backhand side, slide your hands to the two-handed backhand grip and then take your backswing away from the net with your shoulders and body turning in the same direction. Try to keep the racket head below your hands so that when you hit the ball, the strings of your racket brush up the backside of the ball, giving it topspin. Next, take a big step, moving your body forward and your arms up. As you execute the stroke, transfer your weight to your front leg, and then follow through with your racket ending up over your shoulder. From beginning to end, keep your eye on the ball and bend your knees.

That is a *lot* of information to juggle! To do so, you must engage your cognitive control processes—regulated by your prefrontal cortex—to keep the individual steps in mind, and to execute them in the proper order. As a novice, you consciously monitor your performance, second by second; this skill does not come easily, and you must practice until all the critical steps come together. During the learning process, your brain will *chunk* the many individual pieces of your backhand stroke into a single, fluent shot. When you retrieve them, the combined elements of your backhand act as a coherent, integrated group. Weeks, months, and even years later, you will make the shot automatically, without thinking, and then can direct your attention and cognitive control processes to the strategies needed to win the game, set, and match.[6]

Fortunately, the process of motor-skill learning is readily amenable to laboratory study, so Henry became a rich resource. The intriguing evidence from Milner's 1955 mirror-drawing study and my own 1964 study inspired me to examine whether Henry could learn other motor tasks. In 1966, when he was forty years old, I had the opportunity to examine this question more thoroughly. His parents gave their approval for him to check into the MIT Clinical Research Center (CRC) for two weeks of testing—the first of Henry's fifty visits to the CRC over the next thirty-five years. On this trip, the purpose of our tests was to pursue the observation that, in the face of his profound amnesia, Henry could still learn new motor skills. With the prospect of testing Henry on fourteen consecutive days, I recorded his day-to-day progress on three measures of skill learning—rotary pursuit, bimanual tracking, and coordinated tapping.[7]

The apparatus for the first task, *rotary pursuit*, resembled an old-fashioned record turntable with a metal target roughly the size of a quarter located about two inches from the edge. Henry held a stylus between his right thumb and index finger, and I asked him to rest the tip of the stylus on the target. After a few seconds, the disc began spinning, and for twenty seconds he tried his best to keep the stylus in contact with the target as it turned; I recorded the time that the stylus remained on the target, as well as the number of times it left the target. I tested Henry and control participants twice a day for the first two days and once a day for the next five days. Then, I tested them again a week later to see how well they remembered the task without practicing it (see Fig. 10).[8]

Over the seven days of testing, Henry's scores improved, although not as much as the control participants'. A closer look revealed that the number of times he made contact with the target increased with practice; he became more proficient at returning to the target once he lost contact with it. Overall, the control participants stayed on the target longer. Although Henry's gains were not as dramatic as the others, he retained the new motor skill for one week with no additional training. When I tested him on day fourteen, he performed just as well as he had on day seven.[9]

The next week, I trained Henry on a *bimanual tracking* task. The apparatus was an aluminum drum with two narrow, asymmetric tracks painted on it. Henry held a stylus in each hand and placed one on each track. His job was to maintain contact with the tracks as the drum rotated for twenty seconds. This task was especially difficult from a motor-control perspective because Henry's brain had to coordinate the movements of his left and right hands and his eyes, which moved back and forth from one track to the other. The two sides of his brain thus had to interact continuously. I repeated the test three times at increasing speeds of rotation, recording how many seconds Henry and the controls stayed on each track and how many times they fell off. As before, Henry's scores were inferior to the control participants', and he was less consistent, but he again demonstrated a clear improvement from trial to trial in performing this motor skill (see Fig. 11).[10]

Henry's suboptimal performance on rotary pursuit and bimanual tracking was not due to his memory problem; these two tasks depended on quick reaction times. When Henry had more time to respond to a stimulus, he performed just fine. But in general, he tended to do everything slowly. His slow tempo was likely due in part to Phenobarbital, a sedative prescribed for insomnia as well as epilepsy. Other patients who had similar lesions—Scoville's patient D.C., and Penfield and Milner's patients P.B. and F.C.—also took antiseizure medication and moved sluggishly. But despite his slowness, Henry clearly could learn new motor skills and retain that knowledge over long periods of time. We do not know how he would have performed were he taken off his antiseizure medications, which was not an option because doing so would have put his health and safety at risk.[11]

Another motor-learning task, *coordinated tapping*, measured Henry's ability to tap four targets in turn with a stylus, first with each hand separately, and then with the two hands together. The goal of this study was to see whether, with practice, he would speed up and increase the number of times he tapped in the thirty seconds allowed. The apparatus consisted of a black wooden board with two metal circles, side by side, divided into quadrants. Each quadrant was numbered 1, 2, 3, or 4, but the numbers were arranged differently in the two circles. First, Henry held a stylus in his right hand and tapped the circle on the right in the order 1–2–3–4. He then held a stylus in his left hand and tapped the circle on the left in the order 1–2–3–4. I next asked him to tap the two targets simultaneously, which was especially demanding because Henry was required to tap the two 1s simultaneously, the two 2s simultaneously, and so on. He had to coordinate the movements of his left and right hands, and because the location of the numbers differed between the two circles, each hand had a different trajectory to follow. Henry and the control participants performed the task twice, with a forty-minute break between sessions (see Fig. 12).[12]

On this test, Henry scored as well as the control participants, and when I retested him after a break, he was faster than he had been initially. He consolidated the motor memory of the tapping skill, allowing him to demonstrate his learning of this motor behavior forty minutes later. Why was Henry's learning comparable to that of the control group on the tapping task, but not on rotary pursuit and bimanual tracking? A major difference was that the tapping task was self-paced; Henry went at his own speed. On the other two tasks, however, the movement of the apparatus dictated his moves. The rotary pursuit apparatus turned at three different speeds, and the drum on the bimanual tracking apparatus advanced automatically in short steps. On these two tasks, he also had to predict quickly where the target was going, and this need to anticipate the future may have required input from declarative memory.[13]

These early studies with Henry illuminated the distinction between declarative and nondeclarative learning. Declarative knowledge requires medial temporal-lobe structures for its expression, whereas nondeclarative, procedural knowledge is independent of that network. Learning new skills, new procedures, occurs without conscious awareness. When we ride bicycles, play tennis, or ski, we demonstrate our expertise—or lack thereof—through performance. If we try to analyze what we are doing millisecond by millisecond, we may crash, miss a shot, or catch an edge. Similarly, musicians find that their performance falls apart if they try to think about a difficult piece of music note by note; instead, they execute a complex motor sequence without thinking about it. When concert pianist Peter Serkin performs a Mozart concerto with the Boston Symphony Orchestra, his interpretation is driven by his brain's extensive procedural knowledge acquired over years of rigorously practicing that piece; he has integrated the individual key presses into a fluent whole, and performs without conscious reference to individual finger movements.

Before neuroscientists investigated the distinctions between different types of learning, other thinkers in philosophy, computer science, and psychology theorized more abstractly along these lines. The British philosopher Gilbert Ryle wrote about a particular division in his 1949 book *The Concept of Mind*, scolding theorists of the mind for putting too much emphasis on knowledge as the foundation of intelligence and for failing to consider what it means for an individual to understand how to carry out tasks. Ryle termed this difference knowing *that* versus knowing *how*. When we learn a skill, such as a new dance move, we may be unable to articulate the sequence of commands the brain sends to our muscles and the resulting feedback—knowing *that*—but we can show off the new move to our admiring friends—knowing *how*.[14]

Henry's ability to learn new motor skills demonstrated convincingly that the areas that had been excised in his operation—the hippocampus and surrounding structures—were not necessary for learning new motor skills. So of course the next question we wanted to answer was, what critical brain circuits *do* support motor learning? In order to research this question, we focused on non-amnesic patients whose brains were damaged in other ways.

Since the dawn of the twentieth century, scientists have known that two structures, the *striatum* and the *cerebellum*, play important roles in motor control. The striatum includes the *caudate nucleus* and the *putamen*, two collections of neurons under the cortex. They receive signals from above and below—neurons in the cortex and neurons lower in the brain. The striatum receives messages from specific cortical areas and sends signals back to the same areas by way of the thalamus, an area in the center of the brain that integrates sensory and motor activities. As a result, the striatum is well informed about what is going on in the body and in the world, and as such is well qualified to learn difficult motor skills.

The cerebellum, Latin for *little brain*, is a large, complex structure at the back of the brain under the visual cortex. Henry's cerebellum was greatly reduced in size, but we could not tell from his MRI scans exactly where the damage was. This structure is directly connected to the striatum and to several areas in the cortex by means of closed circuits. Because the cerebellum receives information from many parts of the brain and spinal cord, it stands at the frontline of motor control.

Abnormalities in the striatum are responsible for more than twenty disorders, including two progressive brain diseases, Parkinson disease and Huntington disease. Within the striatum, the putamen is most affected in Parkinson disease and the caudate nucleus in Huntington disease.

Parkinson disease is a common affliction with an unknown cause that typically strikes people in their fifties, men more than women. Someone afflicted with this disease often has an expressionless face, slow movement, shaking of the hands, stooped posture, and shuffling steps. In the brain, Parkinson begins with a loss of neurons in the substantia nigra, a bundle of gray matter under the cerebral cortex that normally sends out fibers, which carry the neurotransmitter dopamine up to the striatum. But when cells in the substantia nigra die, as they do in Parkinson, the supply of dopamine transmitted to the putamen is diminished, causing motor abnormalities.[15]

Huntington disease is a rare hereditary disorder caused by neuronal loss in the caudate nucleus, accompanied by cell death in the cortex. The cause is a defect in the HTT gene on chromosome 4. A particular segment of DNA in this gene is repeated up to one hundred twenty times in people with the disease, but only ten to thirty-five times in unaffected people. While the hallmark of Parkinson disease is too little movement, the salient feature of Huntington disease is too much movement. The most striking symptom of Huntington is involuntary, jerky movements of the face, arms, and hips, which makes it seem as if the afflicted person is doing a dance.[16]

Studying Parkinson and Huntington diseases side by side is instructive because the initial damage occurs in different parts of the striatum—the putamen in Parkinson, and the caudate in Huntington—providing complementary evidence about the localization of different skills.

In the early 1990s, to explore the role of the putamen in motor-skill learning, my lab studied mirror tracing in early-stage Parkinson patients. We asked the patients to perform a test similar to the one Milner conducted with Henry. The task was to trace around a six-pointed star as quickly as possible without veering off the track. Because the Parkinson patients had a motor disorder, they took longer to get around the star, traced more slowly, and paused more frequently than control participants. These deficits, which we had anticipated, were measures of motor performance, not motor learning. To see whether Parkinson affected their motor-skill learning, we documented their improvement over three consecutive days of training and then compared their rate of change to that of control participants. A digitizing tablet placed under the star told us precisely where the stylus was from the start to the finish of each trial, millisecond by millisecond. These data allowed us to calculate several different indices of motor-skill learning. These measures were uncontaminated by deficits in motor performance because they focused strictly on each individual's rate of progress, regardless of the performance level at which he or she started.

Although the Parkinson patients improved on all of these measures over the three days of training, their progress was slower than the controls'. On several measures of learning—how long it took them to trace around the star, how long it took them to get back on the path when they veered off, and how much time they spent going backward—the Parkinson patients showed less improvement over the three days than control participants. The difficulty that the patients experienced on this tracing task provided direct evidence that the striatum participates in complex motor-skill learning, giving credence to the idea that Henry enlisted his striatum to learn the motor skill.[17]

The finding that our Parkinson patients were impaired on mirror tracing did not necessarily mean that they would perform poorly on all motor-learning tasks. Many areas throughout the brain support motor behaviors, and it would not make sense for all these areas to be dedicated to a single, all-purpose motor-learning function. The brain is an efficient machine that does not give its component parts redundant tasks. We, therefore, hypothesized that different motor skills engage separate cognitive and neural processes. It was possible that the particular brain circuit within the striatum recruited for mirror tracing would not be necessary to perform a different skill-learning task, such as learning a specific sequence of responses.

To further explore the scope of the skill-learning deficit in Parkinson disease, my lab adapted a sequence-learning procedure in the early 1990s that Mary Jo Nissen and Peter Bullemer first introduced in 1987. Our patients with Parkinson disease sat in front of a computer terminal and saw four small white dots arranged horizontally across the bottom of the screen. A custom-designed keyboard just below held four response buttons corresponding to the four dots. Participants rested their left middle and left index fingers on the two left buttons, and their right middle and right index fingers on the two right buttons. On each trial, a small white square appeared below one of the four dots, and the task was to press, as quickly as possible, the key that corresponded to the location of the square. Unbeknownst to the participants, the squares appeared in a ten-item sequence that repeated ten times on each trial, for a total of one hundred key presses. We knew that if the participants were learning the sequence, their response times would become progressively faster on trials that contained the repeated sequence but would not speed up on other trials in which the sequences were random (see Fig. 13).[18]

Parkinson patients and controls performed the sequence-learning task on two consecutive days. The performance times of the two groups did not differ; the Parkinson patients performed normally. Response times for the repeated sequences decreased during the first day, and participants retained this learning overnight, performing as well at the start of the second day as they had at the end of the first. The Parkinson patients' reduction in response times for repeated sequences indicated that they had acquired procedural knowledge in a normal fashion.

A comparison of the performance by Parkinson patients on the mirror-tracing and sequence-learning tasks, impaired in the former and preserved in the latter, indicates that skill learning is not a homogeneous concept and that different kinds of skill learning have different neural underpinnings. The memory network in the striatum that normally supports the acquisition of the mirror-tracing skill was dysfunctional in our Parkinson group, but a neural circuit that was spared in the same patients mediated normal sequence-specific learning. We then asked, what is this circuit, and is it compromised in other diseases?

Research on Huntington disease provided a clue about the substrate for sequence-specific learning, informing us about the effect of damage to the caudate nucleus on this task. When Nissen administered her sequence-learning task to a group of Huntington patients, they showed impaired learning. Although their motor function was sufficient to perform the test satisfactorily, they were slower and less accurate than the twenty-one control participants. Their deficit was unrelated to cognitive dysfunction. This result tells us that the caudate nucleus plays a critical role in sequence-specific learning.[19]

The experiments in Parkinson and Huntington diseases, examined side by side, illustrate how different pathologies within the striatum can exert distinct effects on sequence learning: Parkinson patients are normal, while Huntington patients are deficient. This dissociation suggests that the caudate nucleus, affected early in Huntington, is a critical substrate for sequence learning, and that the putamen, affected early in Parkinson, is not.

Brain areas beyond the striatum are also engaged in motor learning. Since the late 1960s, neuroscientists have gained another perspective on the brain localization of skill acquisition from studying animals and patients with abnormalities of the cerebellum. Their symptoms include poor coordination, slow movement, tremor, and slurred speech. People who are heavily intoxicated demonstrate these symptoms, as did Henry, minus the tremor. One group, patients with cerebellar degeneration, are impaired at sequence learning, but several studies show that their fundamental deficit may differ from and be more severe than that in Parkinson patients. Patients with cerebellar degeneration were also slower and less accurate than control participants when tracking a simple geometric pattern that they saw in mirror-reversed vision, similar to the conditions under which Henry successfully traced around the star. In 1962, we learned from Henry that medial temporal-lobe structures are not necessary for learning the mirror-tracing skill, and thirty years later we learned that the cerebellum *is* necessary for that kind of learning.[20]

The deficit in mirror tracing in cerebellar disease reflected the patients' inability to use the feedback they received during the test to guide their movements. Although they could see the visual display and feel the changing position of their arms and hands, they could not convert this information into new commands to activate their muscles. They could not overcome their ingrained responses. This difficulty is not specific to a single task but essentially a general failure to integrate input from the senses with commands to the muscles. Consider typing on a keyboard as an example. When we perform this skill, we receive information from several sources—the feel of the keys on our fingertips, the position and movement of our fingers and hands, and the visual image of our hands and the document on the computer screen. When we type, our brains automatically combine all these inputs, telling our fingers how to move to hit the right keys in the right order and with sufficient force to make the desired letters appear on the screen. Healthy people have no difficulty acquiring this complex motor skill if they practice.

A striking example of coordinating sensory and motor circuits is prism adaptation. To perform this task, participants wear glasses with prisms that bend the light a few degrees to the left or right, making objects appear displaced to the left or right of their true locations. Before donning the prisms, participants practice pointing to a target with normal vision. Once they are proficient, the experimenter asks them to wear the prism glasses, thereby changing the visual environment in which the target is presented. If the prism glasses displace the target slightly to the left, participants at first point to the right of the target. But after practicing for a few minutes, they update their movements and eventually reach the target. When they remove the prisms and point again, they show an aftereffect of adaptation—pointing in the opposite direction, indicating that they had adapted to the altered visual information.

Research in the late 1990s helped uncover the brain circuits required for the adaptive process. To pinpoint the specific area critical for accommodating changes in the visual environment, neuroscientists administered a prism-adaptation task to patients with disorders of the cerebellum. In a 1996 experiment, participants threw balls at a target under three conditions: before donning the prisms, while they wore them, and immediately after they removed them. The researchers assessed learning during the third condition. Because the prisms made the target appear to the left of where it really was, participants initially threw the balls to the left side of the target. With practice, they gradually threw more and more to the right, and the impact points moved back toward the center of the target. After removing the prisms, the control participants continued throwing to the right side of the target, as if they were still wearing the prisms, indicating that they had adapted to the visual shift. This *negative aftereffect* is the measure of learning. The patients with cerebellar disorders did not show a negative aftereffect, convincing proof that their brains had not retained the altered map created by the prisms. This experiment shows that the cerebellum integrates two kinds of information, perceptual and motor, to accommodate vicissitudes in the visual world.[21]

Remarkably, when we tested him in the mid 1990s, Henry showed normal adaptation to prisms, despite having marked cerebellar atrophy. The prism-adaptation task was ideal for testing the effects of his cerebellar damage on a kind of nondeclarative learning that grows from interactions between brain circuits specialized for visual perception and for movement. Our experiment tested whether Henry's motor system could adapt to a situation in which prisms displaced everything in his work-space eleven degrees to the left. To accomplish this visual shift, we asked him to wear glass prisms set in a pair of laboratory goggles. His task was to point quickly with his right index finger to a vertical line at arm's length in three situations: a baseline condition without prisms, an exposure condition with prisms, and a post-exposure condition without prisms. In each condition, Henry pointed to nine different targets, one straight ahead and four to each side. We presented each target four times in random order. For each trial, we recorded the position of Henry's finger and then determined how far it was from the target. As in other prism-adaptation experiments, the measure of learning was the amount of negative aftereffect in the post-exposure condition—how far the point he touched deviated from the target.

Henry performed just like the ten control participants. In the exposure condition, he could clearly see that he was pointing far to the left of the target and gradually shifted his pointing to the right to hit the target dead-on. When the prisms were removed, he continued to point to the right of each target as if the prisms were still in place—clear evidence of a normal aftereffect. During the experiment, sensory and motor circuits in Henry's brain interacted successfully to accomplish this nondeclarative learning.

Although we do not yet know what residual cerebellar function supported Henry's good performance, we hope to understand these results more fully as we examine his postmortem brain and identify the specific cerebellar circuits that were left intact. Of particular interest are structures that transmit information to the cerebellum—the deep cerebellar nuclei—which if spared may have provided the necessary machinery for prism adaptation. Figuring out the anatomical substrate for prism adaptation will be a noteworthy accomplishment.

Our investigations of motor-skill learning contrasted Henry's performance with that of other patients who had damage to areas outside the medial temporal lobes. We learned that motor-skill learning and declarative memory are assigned to different compartments in the brain. The hippocampal region contains the circuits critical for recalling and recognizing facts and events, but not for learning new motor skills. In contrast, circuits in the caudate nucleus, putamen, and cerebellum are necessary for motor-skill learning, but not for retrieving facts and events.

Although Henry could acquire new skills in the laboratory, this ability did not bring much benefit to his everyday life, except for mastering the walker. The symptoms caused by the damage to his cerebellum, on top of his epilepsy, were not conducive to dancing or learning new sports. He did play croquet, but we do not know whether his game improved with practice.

In addition to studying brain-damaged patients to deconstruct the neural architecture of motor skills, scientists have proposed theoretical models to explain how the brain learns and then performs these tasks. In 1994, neuroscientists Reza Shadmehr and Ferdinando Mussa-Ivaldi at MIT provided a major breakthrough in understanding motor memory, introducing the idea that when the body makes reaching movements, the motor-control system adapts to unanticipated changes in the environment. The brain accomplishes this feat by constructing an *internal model* that, with experience, estimates the forces in the environment—the pushing and pulling. The concept of an internal model has become a popular explanation for how the brain represents and modifies learned skills.[22]

To understand internal models, imagine that you are thirsty; you pour a glass of water, grasp it, bring it to your lips, and drink. This simple action, which you have carried out many times in many different places, is not as straightforward as it seems. Before you move your arm, your brain receives and processes basic information about the glass: its shape, how heavy it is likely to be, where it is located, and where your hand is. The problem for your brain is to translate the location of the glass on the table and your goal, grasping the glass, into the pattern of muscle activity necessary to bring the glass to your lips. We execute this kind of motor command constantly as we move through our day—brushing our teeth, using a knife and fork, driving a car, browsing the web. Over our lifetimes, we interact with innumerable different objects in a vast set of environments, and each time our brains must transform information from our senses into movement, and luckily can readily adapt to changes from one situation to another.

Internal models represent circuits in the brain that process the relation between the motion of the hand and the motor commands. For example, an *inverse model* embodies the relation between the *desired* motion of the hand and the motor outputs required to achieve this motion. This sort of internal model is a major component of a system that can guide your hand to grasp the glass. Another type of internal model, a *forward model*, allows the brain to predict the likely outcomes of a motor command and choose those necessary to perform specific motor tasks successfully—in our example, having a drink of water. In 1998, a computational neuroscientist in Japan, in collaboration with a colleague in London, adopted the idea of internal models and proposed that acquiring a new motor skill depends on establishing such internal models for motor-task performance. Motor learning is a process of translating the spatial characteristics of the target or the goal of the movement into an appropriate pattern of muscle activations.[23]

The computational neuroscientists proposed that the two kinds of internal models work cooperatively to track what we are actually doing and create a mental picture of the movement we want to achieve. One model registers the link between motor outputs—reaching for and grasping the glass—and the ensuing sensory inputs—the glass and the position and velocity of your arm. This model makes step-by-step predictions about the next position and velocity of your arm, given the present state of your arm and the reaching command (go to the glass). The other model provides the actual motor command needed to grasp the glass.[24]

When these two internal models interact, the brain compares the actual state of the arm with the desired state of the arm; the discrepancy provides critical information about errors in performance. Error messages facilitate learning by indicating how to adapt the movement to reduce errors and to achieve the desired goal. The brain can switch from one internal model to another based on contextual information—new location of the glass—or on error information—sensorimotor feedback about accuracy. This switching mechanism guarantees flexible adaptation in constantly and rapidly changing environments.[25]

When Henry was learning to trace around a star while seeing the pattern, the stylus, and his hand only in mirror-reversed view, he was building up new internal models in his brain depicting the relation between what he saw and how he moved his pencil. These novel internal models had dedicated circuits in his brain, so they did not interfere with all the other motor behaviors he had learned previously. In everyday life, we accumulate many such internal models to build an enormous repertoire of complex motor behaviors.

Based on evidence from computational modeling, cognitive science, and neurophysiology, researchers in Kyoto, Japan, predicted that internal models are predominately created and stored in the cerebellum. This large, complicated structure is qualified for this assignment because it has the physiological capability to compare the desired movement with the actual movement, and then to use this difference—an error signal—to guide the next movement.

In 2007, when the Japanese researchers tested this hypothesis with functional MRI, they obtained the first physiological evidence that internal models are formed in the cerebellum. These scientists conducted a series of experiments in which participants executed a tracking task, moving a computer mouse to keep the cursor on a target that moved randomly on a computer screen. In the baseline condition, the mouse was in the normal orientation, but in the test condition, it was rotated 120 degrees, changing the relation between the mouse and cursor and forcing participants to learn how to control the mouse in a new way. Training occurred over eleven sessions, with functional MRI scanning on the odd-numbered sessions to capture the neural activity associated with the learning process from beginning to end.

During testing, the researchers found two separate regions of activity in the cerebellum. One was an error-related region where neural activity decreased as learning progressed and tracking became more accurate.

The second was unrelated to tracking errors; instead, it was an internal-model-related region where activity continued to appear late in training, and seemed to be the site where an enduring internal model of the new tracking skill was stored. The neural activation for this motor-learning task occurred in many areas on both sides of the cerebellum, some of which receive helpful information from the frontal and parietal cortices about planning, strategy, and reaching movements.[26]

Considering the evidence about the critical role of the cerebellum in motor-skill learning, it was surprising to me that Henry, whose cerebellum was severely damaged by his medicine, did as well as he did on mirror tracing, rotary pursuit, and bimanual tracking. My initial studies had been limited because they provided only crude measures of Henry's performance—how many errors he made and how long it took him to complete a task. I sought a deeper understanding of how Henry's brain controlled his movements throughout the skill-learning process. In 1998, an exciting and fruitful collaboration with Shadmehr, a researcher at Johns Hopkins University, enabled us to examine Henry's motor memory processes in greater detail during the course of learning. Shadmehr had been a postdoctoral fellow in my department, and I was impressed by his research and expertise in the field of motor control. I, therefore, invited him and two of his students to MIT to conduct a skill-learning experiment.

The motivation for our experiment was a 1996 study demonstrating that *consolidation* of the motor-learning experience continues after learning, an insight that came from examining motor-memory consolidation in healthy young adults. When the participants performed a motor skill they had practiced in a previous session, they immediately performed better than they had at the end of the last training trial, indicating that the memory had improved over the intervening time. This gain, however, was disrupted when participants were instructed to learn a second motor task right after the first. Their consolidation of the first task was disturbed because of interference from the second task. In contrast, no disruption occurred if four hours elapsed between learning the two motor skills. The study suggests that consolidation in motor memory happens rapidly—over a period of just four hours after practice, the memory of a new skill was transformed from an initial fragile state to a more solid state. This rapid time-course contrasts with the consolidation of declarative memories, which may require years.[27]

This discovery made in the laboratory often carries over into personal experiences. One of my editors tells her ski instructors that she can learn only one new skill per lesson. If they try to teach her two or more, she does not learn anything at all because the consolidation of one new skill is interrupted by switching to another.

Shadmehr's 1996 experiment in healthy young adults raised important questions about skill learning: does the consolidation of motor memories require that the participants remember declarative information about the task? Does the medial temporal-lobe need to function normally for the interference effect to occur? Because declarative memory was operational in the young adults we studied, the answers had to come from a memory-impaired participant. Studying Henry, whose declarative memory was decimated, could tell us definitively whether this source of knowledge mattered. Our study was the first to examine the process of interference associated with motor memories in amnesic patients. If declarative memory played no role in the interference of motor memories after practice, then the consequences of learning multiple motor skills should have been the same for Henry and the control participants.[28]

During a two-day experiment, we studied Henry's ability to learn a novel motor skill. The task was not a video game, but it resembled the Wii game *Link's Crossbow Training*, in which players shoot at bull's-eyes as they pop up on the screen. Initially, the targets are stationary, and when the player hits one, it explodes; as the game advances, the bull's-eyes move, making the task more difficult. In our experiment with Henry, the targets were always stationary. After he became proficient at firing straight at the targets, we introduced an unexpected change by mechanically jolting his arm as he moved, thus throwing him off-course. We wanted to know whether with training, his moves to the target would become straight again (see Fig. 14).

The apparatus for the task was a mechanical arm with a video monitor located just above it. When the researchers first sat Henry in front

1. Four lobes of the cerebral cortex

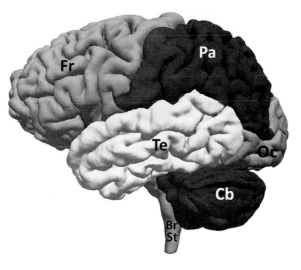

MRI scan of a healthy forty-one year old man, showing his whole brain viewed from his left side, with the four lobes of the cerebral cortex delineated. The frontal lobe (Fr) modulates basic motor functions and cognitive control processes (setting goals, making decisions, solving problems); the temporal lobe (Te) complex visual and auditory processes, memory, language, and emotion; the parietal lobe (Pa) touch, pain, and other body sensations as well as spatial ability and language; and the occipital lobe (Oc) basic visual processes. The peaks in the cortex are the gyri and the valleys between them are the sulci. The cerebellum (Cb) is specialized for balance and movement coordination. In Henry's brain, this structure was badly shrunken as a side effect of Dilantin. The brainstem (BrSt) connects the spinal cord to the rest of the brain. It is the entryway for information from the senses, and circuits in the brain stem control vital bodily functions, such as heart rate, blood pressure, respiration, and level of consciousness.

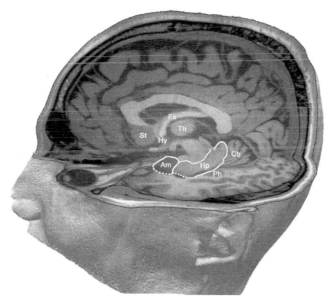

2a. Medial temporal-lobe structures

The same healthy brain showing the medial temporal-lobe structures that were removed from Henry's brain. The amygdala (Am, dark gray outlined in white) and the head and body of the hippocampus (Hp, light gray outlined in white). The tail of the hippocampus has been removed for simplicity but would normally continue upward forming the fornix (Fx), which arches forward to the mammillary bodies of the hypothalamus (Hy). The back part of the parahippocampal cortex (Ph) was spared in Henry, but the front part was removed. Other structures shown are the cerebellum (Cb), the striatum (St), an area engaged in motor control and motor learning, and the thalamus (Th), a structure where information from the eyes, ears, and skin is relayed to pathways that carry the input to the cortex.

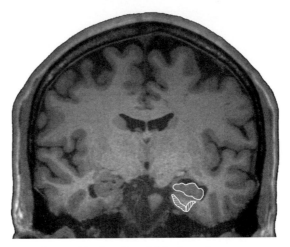

2b. The hippocampus and entorhinal and perirhinal cortices
A different view of the same brain, as if we are face to face with the person. The gray matter follows the cortical ribbon along the entire cerebral cortex, while the white matter lies beneath the gray matter. The hippocampus and entorhinal and perirhinal cortices reside in the medial temporal lobe seen in the lower right of the image. The hippocampus is outlined with a solid white line, the entorhinal cortex is shown with horizontal hatched lines, and the perirhinal cortex is shown with vertical hatched lines. Henry's operation removed all three structures on both sides of his brain.

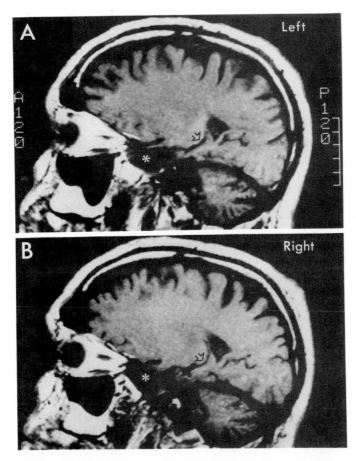

3. Henry's MRI
These MRI images collected in 1992 show Henry's lesions on the left and right sides of his brain. The *asterisk* indicates the missing portion of the medial temporal lobe structures. The remaining portion of the hippocampal formation is indicated with an *open arrow*. We can see approximately two centimeters (.79 in.) of remaining hippocampal formation on each side. The enlarged spaces between the leaves of the cerebellum are evidence of substantial cerebellar degeneration.

4. Mooney Face-perception Test
An item from the Mooney Face-perception Test. When Henry was forty, we asked him to state the sex and approximate age of each person depicted in the test stimuli. He scored higher than the control participants, indicating that his visual perception was intact.

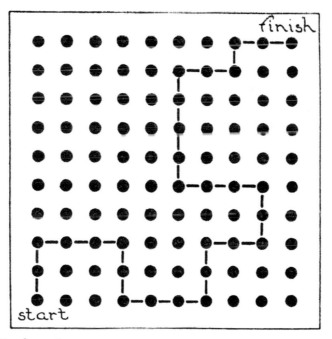

5a. Visual stepping-stone maze
The visual stepping-stone maze consists of black dots, which are metal bolt heads on a wooden base. Henry's task was to discover and remember the correct route (indicated by the black line). As he proceeded from bolt head to bolt head, the click of an error counter signaled each mistake. During three days of training when Henry completed two hundred fifteen trials, he did not succeed in reducing his number of errors, revealing a deficit in his declarative memory.

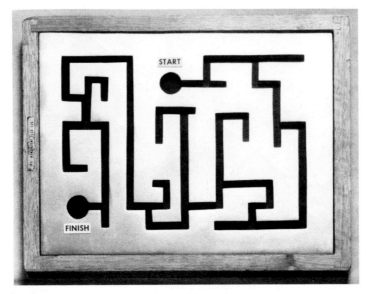

5b. Tactual stylus maze

The tactual stylus maze rested inside a wooden frame. Henry sat on one side, where a black curtain covered the frame to prevent him from seeing the maze. I sat on the opposite side, which was open so I could observe his hand, the stylus, and the maze as Henry advanced through it. I instructed him to move the stylus along the paths to find the correct route from the start to the finish, and each time he entered a blind alley, I rang a bell signaling that he should back up and try another path. On four consecutive days, Henry completed two sessions of ten trials each but without decreasing his error scores, indicating a failure of his declarative memory.

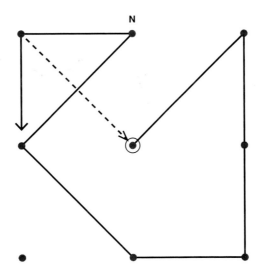

6. Route-finding task

One of fifteen large maps used to test Henry's *spatial ability*. The map depicted the nine red circles on the floor of the testing room. A path from dot to dot was drawn in heavy black lines, with a circle around the starting point and an arrowhead at the end. An N, designating north, was marked on each map, and a large red N was fixed on a wall of the room. Henry's task was to walk from dot to dot along the path that corresponded to the one on the map. He walked patiently from dot to dot, but was usually unable to follow the path indicated on the map, and his performance did not improve with repeated testing using the same maps. Here, the dotted line shows where he went off course.

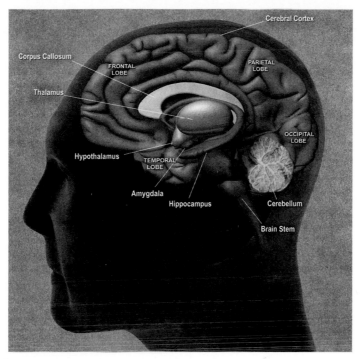

7. Limbic system

Although the idea that a single brain system supports emotion is no longer tenable, the term limbic system is still used to designate a group of interconnected structures that play a role in the appreciation and expression of emotion: the hypothalamus, thalamus, amygdala, cingulate cortex, and orbital frontal cortex. The orbital frontal cortex is the area above the eye, and the cingulate cortex is the gyrus just above the corpus callosum. Because the amygdala and hippocampus are heavily interconnected, our emotions can influence memory formation, but the hippocampus itself does not modulate emotions.

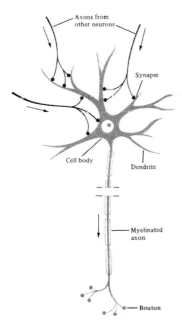

8. The typical neuron

Our brains contain billions of neurons that are constantly talking to each other. The typical neuron has several parts. Each neuron has a *dendritic tree* that receives thousands of signals from other neurons. This information is processed in the *cell body* and from there, it travels along the *axon* for transfer to other neurons. The point where neurons contact each other is the *synapse*.

9. Mirror tracing task

Milner's trailblazing discovery in 1962 showed for the first time that amnesia spares some kinds of learning. She gave Henry the mirror tracing task, which required him to trace around a star—a challenging enterprise because he could only see the star, his right hand, and the pencil in a mirror. Even so, his performance improved measurably during three days of practice. At the same time, he had no conscious knowledge of this experience or his accomplishment, suggesting that the brain houses two different kinds of long-term memory, one in which he succeeded (nondeclarative) and another in which he failed (declarative). Decades later, my lab repeated the mirror tracing experiment and showed that Henry retained the skill when we retested him almost a year after our first test session.

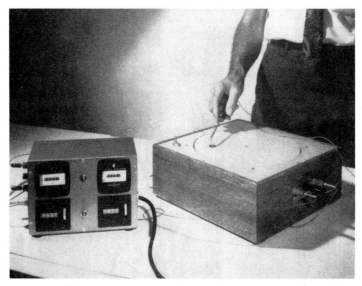

10. Rotary pursuit task

To begin the rotary pursuit task, I asked Henry to rest the tip of the stylus on the target. The disc began spinning, and for twenty seconds he tried his best to keep the stylus in contact with the target; I recorded the time that the stylus remained on the target, as well as the number of times it left. Over the seven days of testing, Henry's scores improved, although not as much as the control participants'. Following an additional week without practice, he retained this skill.

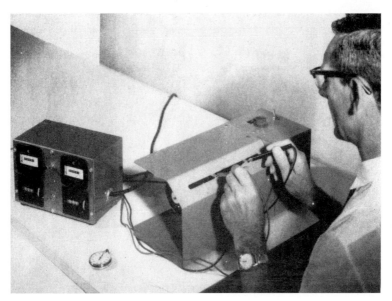

11. Bimanual tracking task

Henry's job on the bimanual tracking task was to maintain contact with the tracks as the drum rotated for twenty seconds. This task was especially difficult from a motor-control perspective because his brain had to coordinate the movements of his left and right hands and his eyes, which moved back and forth from one track to the other. Although Henry's scores were inferior to the control participants', and he was less consistent, he again demonstrated a clear improvement from trial to trial on this motor learning task.

12. Coordinated tapping task

To perform the coordinated tapping task, Henry held a stylus in his right hand and tapped the circle on the right in the order 1–2–3–4. Then, with the stylus in his left hand, he tapped the circle on the left in the order 1–2–3–4. I next asked him to tap the two targets simultaneously, which was especially demanding because he had to tap the two 1s simultaneously, the two 2s simultaneously, and so on. He had to coordinate the movements of his left and right hands, and because the location of the numbers differed between the two circles, each hand had its own trajectory to follow. On this self-paced task, Henry scored as well as the control participants, and when I retested him after a break, he was faster than he had been initially.

13. Sequence-learning task

For the sequence-learning task, we asked our Parkinson disease patients to sit in front of a computer terminal and view four small white dots arranged horizontally across the bottom of the screen. On each trial, a small white square appeared below one of the four dots, and the participant's task was to press, as quickly as possible, the key that corresponded to the location of the square. Unbeknownst to the patients and controls, the squares appeared in a ten-item sequence that repeated ten times on each trial,

for a total of one hundred key presses. We found that the early stage Parkinson patients and controls learned the sequence: their response times became progressively faster on trials that contained the repeated sequence but did not speed up on other trials in which the sequences were random. In contrast, the patients with Huntington disease did not show this nondeclarative learning.

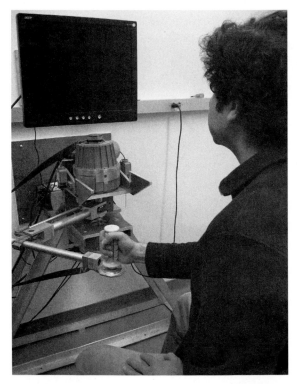

14. Reaching task

For the reaching task, we showed Henry individual targets on the screen and asked him to use the mechanical arm to move the cursor to those locations as quickly as possible. His goal was to reach the target within one second, and each time he succeeded, the target exploded. After Henry spent a few minutes moving the cursor to the targets, we changed the procedure without warning; the mechanical arm imposed a force on his hand, throwing his movements off-course to one side. With practice, however, he altered his motor commands to compensate for the force, and was again able to quickly move his hand in a straight line to the targets. We knew that Henry had learned to compensate for the force because when we suddenly removed the force, his movements had large errors, just like the pattern of errors he had made early in training but reversed.

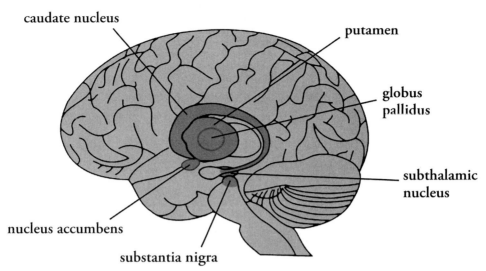

caudate nucleus

putamen

globus pallidus

subthalamic nucleus

nucleus accumbens

substantia nigra

15. Basal ganglia

The basal ganglia are a distributed set of interconnected circuits that work in collaboration with the cortex to achieve the control of posture, movement, and unconscious learning. The key basal ganglia structures are the striatum (caudate nucleas and putamen), nucleus accumbens, globus pallidus, subthalamic nucleus, and substantia nigra. Information flows in a loop from areas in the frontal and parietal lobes through the basal ganglia and thalamus and then back to the frontal cortex.

16. Eyeblink conditioning task

For the eyeblink conditioning task, Henry sat in a comfortable chair and wore a headband that held an air-puff jet and a monitor to record his eyeblinks. We gave him these instructions: "Please make yourself comfortable and relax. From time to time, you will hear some tones and feel a mild puff of air in your eye. If you feel like blinking, please do so. Just let your natural reactions take over." Over an eight-week period, Henry performed two kinds of conditioning tasks: delay conditioning and trace conditioning. Although his performance was inferior to that of the control participant, he did produce conditioned responses in both the delay and trace procedures—evidence of nondeclarative learning.

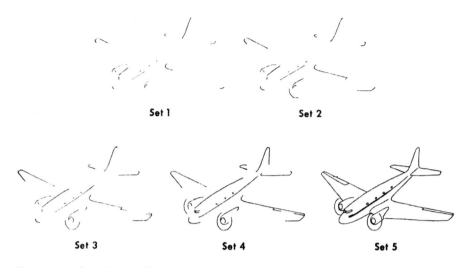

Set 1   Set 2

Set 3   Set 4   Set 5

17. Gollin Incomplete Figures Test

The Gollin Incomplete Figures Test is a measure of perceptual learning. Henry's task entailed viewing simple line drawings of twenty common objects and animals, such as an airplane and a duck. The test began with a very sketchy representation containing a few parts of each object, making it almost impossible to interpret, and ended with a complete, recognizable picture. Henry first saw the most fragmented set, one drawing at a time, each for about a second, and said what he thought the drawing might represent. He then saw progressively complete sets of pictures until he could name all twenty objects. He completed the test without errors after four trials, and remarkably, his accuracy was a bit better than that of the ten control participants. An hour later and without warning, Henry saw the same set of pictures again and identified the fragments in fewer trials. He had learned a perceptual skill without explicit knowledge, and it stuck—solidly stored in the preserved cortical areas of his brain.

18. Pattern priming

The dot patterns for pattern priming are in column one. Examples of the target figures that Henry copied are in column two. Some other figures that could be drawn are shown in the remaining columns. After Henry copied a target figure onto a dot pattern, he was later likely to draw that figure onto a dot pattern when asked to draw whatever he wanted. On three different forms of the pattern-priming test administered on three separate testing occasions, Henry showed a normal magnitude of priming, attesting to his intact nondeclarative memory.

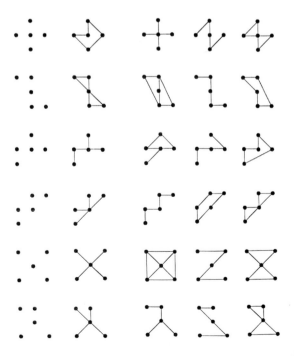

19a. William Beecher Scoville

19b. Brenda Milner,
circa 1957

20. Five-year-old Henry and his parents

21. Henry the animal lover

22. Henry before his operation

23. Henry, 1958

24. Henry with his parents after his operation

25. Henry, 1975

26. Henry ready for testing at MIT, 1986

27. Henry at Bickford Health Care Center

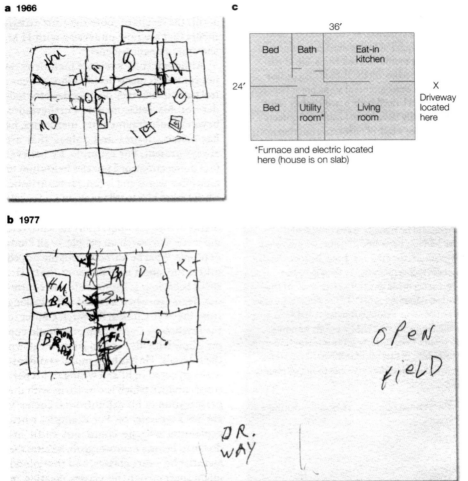

a 1966

b 1977

c

	36′		
Bed	Bath	Eat-in kitchen	
Bed	Utility room*	Living room	

24′

X
Driveway located here

*Furnace and electric located here (house is on slab)

28. Henry's drawing of the floor plan of his home

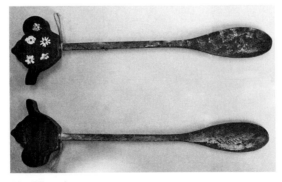

29. Henry's spoon

of the arm, he, like all inexperienced volunteers, sat quietly without touching the machine. They asked him to grasp the handle of the mechanical arm and move it around a bit to get used to it. At first, he kept his gaze on his hand as he moved the handle, but then they told him to look at the monitor, where a cursor was present. After Henry moved the cursor around for a minute or so, the researchers illuminated a target at the center of the screen and asked him to move the cursor to that location. They then showed him other individual targets and asked him to move the cursor to those locations as quickly as possible. His goal was to reach each target within one second. Each time he succeeded, the target exploded.

For Henry, the target explosion triggered childhood memories of going hunting for small game. As he performed the task and earned numerous explosions, he described these cherished memories in detail—the type of gun he used, the porch in the rear of his childhood home, the terrain of the woods in his backyard, and the kinds of birds he hunted. Smiling and excited, he repeated these facts many times during the two-day experiment. It was an emotionally upbeat experience for him.[29]

After Henry spent a few minutes moving the cursor to the targets, we changed the procedure without warning; the mechanical arm imposed a force on his hand, throwing his movements off-course to one side. So, instead of moving to the target in a straight line, his hand swerved on the way to the target. With practice, however, he altered his motor commands to compensate for the force, and was again able to quickly move his hand in a straight line to the targets, consistently achieving the timing goal of 1.2 seconds or less. His brain constructed an internal model of the skill that allowed him to estimate the force in the mechanical arm and counterbalance its effects. That he learned to compensate for the force was evident: when the researchers suddenly removed the force, his movements had large errors, just like the pattern of errors he had made early in training but reversed. At the end of the session, the researchers politely thanked Henry for his time, and he left to have lunch.[30]

Four hours later, when Henry returned to the testing room, he had forgotten entirely about the apparatus and the experiment. The researchers pushed the mechanical arm aside and asked Henry to sit down. He sat down, and then something interesting and unexpected happened. In contrast to the first time he encountered the equipment, this time he voluntarily reached out and grabbed the handle, brought it toward him, and looked at the video monitor in anticipation of seeing a target. Clearly, despite having no conscious recollection of having performed the task before, some part of Henry's brain understood that the contraption was a tool that enabled him to move a cursor on the monitor. When a target was presented, he showed strong aftereffects of the previous training. Because his brain expected the mechanical arm to perturb his movements as it had before, it generated motor commands to compensate for these forces, and he moved the handle to the target as if the force were still in place. The motor memory was much more than knowing how to manipulate a tool; it included information about the rewarding nature of the tool's purpose. In essence, "when I move the handle quickly, something fun will happen." The sight and touch of the mechanical arm were sufficient to encourage a motor act that Henry expected to be rewarding. If in the first session, use of the mechanical arm had been paired with a shock or another noxious stimulus, then Henry likely would have been reluctant to use the device again.[31]

Henry's performance on the reaching task demonstrated that his brain had gained three important insights, all without conscious awareness and without the use of his medial temporal lobes. First, during the initial training session, he learned to use a novel tool to achieve a specified goal, both without and with the disrupting force. Second, when tested hours later, the sight of the tool was sufficient to produce voluntary use, suggesting that Henry had learned and stored the potential rewards associated with the use of the tool—the challenge of earning the explosion. Third, seeing and holding the tool were sufficient to allow him to unconciously recall both the purpose of the tool and the motor commands needed to achieve that purpose—even though the same visual and tactual information were *insufficient* to evoke a conscious memory that he had previously trained on the task.[32]

Unlike the earlier motor-learning experiments using rotary pursuit, bimanual tracking, and tapping, the mechanical-arm experiment allowed us to examine two properties of motor control separately: *kinematics* and *dynamics*. Kinematics refers to the speed of motion, changes in speed, and direction of motion, while dynamics refers to the effect of forces on motion. Although Henry experienced considerable difficulty learning the kinematics of the task, he finally learned that he had to move the arm away from himself to make the cursor go up on the screen and toward himself to make it go down. He was also able to compensate for the imposed force (the dynamics) and move his hand in a straight line to the targets. The goal of our experiment was to see whether Henry's impaired declarative memory would have any effect on the acquisition of these complex motor memories; remarkably, it did not. Just like the control participants, his brain could build new internal models to support the learning of this motor skill.

Milner's trailblazing discovery in 1962 that Henry could learn a new motor skill was a tremendous advance, providing a new way to understand how we acquire and retain nondeclarative memories. Since then, researchers have devised thousands of experiments to shed light on the cognitive and neural processes that support this kind of memory. Currently, experiments focus on the cellular and molecular mechanisms of neuroplasticity in the brain circuits that underlie skill learning. Accrued knowledge from these findings may point the way toward treatments for such diseases as Huntington and Parkinson.

Because movement is a fundamental requirement for interacting with the world, the performance of motor skills is critical to our independence. One mystery of motor skills is how we are able to execute them so quickly and with so little thought. When we are first learning a new skill, we need to put a lot of concentration and effort into it in the form of executive control. Over time, skills that we acquire become increasingly automatic; they require much less mental exertion. Researchers have studied how new motor skills become automatic, and through the use of brain-imaging techniques, we can see how brain activity changes as people master skills.

A big gap in scientific knowledge remains, however. How do motor mechanisms in different parts of the brain—the primary motor cortex, the striatum, and the cerebellum—coordinate their individual contributions to achieve the complex enterprise of motor learning necessary in our constantly changing world? Funcional MRI holds great promise for dissecting the individual processes that govern different kinds of skill learning and for documenting when and how the various brain networks work together. These studies indicate that many cortical areas are recruited during motor learning, suggesting that a broad network of motor and non-motor areas supports skill acquisition (see Fig. 15).

Any athletic skill—dribbling a soccer ball, sinking a free throw, serving an ace—requires extensive training. Increasingly improved performance is coupled with changes in the brain. In 1998, a team of neuroscientists at the National Institute of Mental Health set out to examine the neural changes that occur with continued practice, and to discover how much practice is needed to bring about detectable changes during skill acquisition. They chose the primary motor cortex, a strip at the back of the frontal lobe that sends out neural codes for movements, as their area of interest because it controls voluntary movement and also supports motor learning. The researchers asked healthy adults to practice making sequences of finger movements for several weeks. The participants touched their thumb to the other four fingers one at a time in a particular order: pinky, index finger, ring finger, middle finger, and pinky. They practiced this sequence for ten to twenty minutes every day for five weeks, and as the weeks passed, they completed more sequences and made fewer errors in each thirty-second test.[33]

To capture what was happening in their brains, the researchers conducted a functional MRI study once a week, in which the participants performed the sequence inside an MRI scanner. The resulting MRI images showed activation in the hand area of their primary motor cortex that expanded as the participants' skills improved, and the change lasted for several months. This finding provided evidence that practicing a motor skill encourages additional motor neurons to be active and incorporates them in a focal brain circuit that represents the trained-motor sequence. This indisputable evidence of neural plasticity in adult brains may represent the kind of modification that is responsible for motor-skill learning. The main function of the primary motor cortex is to tell our muscles what to do, but in addition, neuronal firing in this area during motor learning can change synaptic strength—the ability of one cell to excite its synaptic partner cells—thereby promoting memory consolidation. The neural circuits within the primary motor cortex are adaptable on a moment-to-moment basis during the acquisition, consolidation, and retrieval of motor skills.[34]

We demonstrated in the lab that Henry could acquire several motor skills, such as mirror tracing, rotary pursuit, and bimanual tracking. His normal primary-motor cortex likely played a role in his capacity to acquire these new motor skills and to use his walker deftly in everyday life. But it is also probable that helpful changes occurred in other areas in Henry's brain—some dedicated to motor function and others to cognitive processes.

Motor learning usually occurs slowly over many practice sessions, and the complex mechanisms that support skill acquisition change as learning progresses. Training-induced plasticity can be seen in expansions of both gray matter, the cell bodies of neurons, and white matter, the fiber tracts that connect different groups of cells. Initially, the primary-motor cortex and adjacent motor areas are called online, with increased activity in the prefrontal and parietal cortices and the cerebellum as well. Later, when the skilled movements become more automatic, learning still engages the primary-motor cortex and, in addition, the striatum and cerebellum. Movement representations expand in the motor cortex and other cortical areas that are specialized for planning, perceiving movement, controlling eye movements, and calculating spatial relations. These areas work together to achieve motor-memory formation. Multiple circuits throughout the brain are engaged in motor-skill learning, but, as Henry showed us, those in the medial temporal lobes are unnecessary.[35]

Modern brain-imaging tools have enabled us to observe in healthy individuals what the critical circuits are doing during the course of practice.

Researchers wanted to know what areas are active as people advance from novice to expert on a particular task. In 2005, neuroscientists used functional MRI to show that when participants received extensive training on a sequence-learning task (Nissen and Bullemer's task, described earlier), their brain activity in the novice stage differed from that in the later, automatic stage. Initially, areas in the prefrontal cortex and a deep motor area, the caudate nucleus, were highly active, but this activity decreased when performance became automatic with practice, suggesting less reliance on cognitive control processes. The finding that the striatum (caudate and putamen) plays a key role in the acquisition of motor-sequence knowledge is consistent with the finding of motor-learning deficits in Parkinson disease and Huntington disease, both of which damage the striatum.[36]

Starting with the idea that learning occurs gradually from one training session to the next, two neuroscientists at Concordia University in Montreal conducted an ambitious study in 2010 to document changes in brain activity across five consecutive days of skill acquisition. Scanning participants at every training session was unnecessary, so they performed the same motor-learning task inside the scanner on days one, two, and five, and outside the scanner on days three and four. The researchers discovered that, as performance improved, several motor areas that were initially active became less active. These decreases may have occurred because the brain was paying less attention to repeated stimuli and no longer needed to correct errors as learning progressed. At the same time, small areas within the primary-motor cortex and the cerebellum showed increases in activity with improvement.[37]

These pockets of increased activity within a network of overall decline in activity could represent the areas where motor memories are ultimately stored. The researchers speculated that separate populations of neurons within the primary-motor cortex encode and express different facets of motor-sequence learning. One population of neurons, which is activated by performance errors, is dedicated to *rapid learning*; it talks to a declarative memory network. The other population, which shows resistance to forgetting, is specialized for *gradual learning*; these neurons talk to a network dedicated to learning procedures—the *how* to do it. These two populations of neurons work cooperatively.

We now have compelling evidence that the evolution of a complex skill from novice to expert is not a single process. Different timescales operate in motor memory, and their contributions change over time. The ability to tease apart neutrally distinct processes helped us to understand Henry's performance on the reaching task when he had to compensate for the added force in the mechanical arm. Although he could retain the skill, his rate of learning lagged behind that of the controls. The results of the 2010 functional MRI experiment lead me to speculate that Henry's slow progress could be blamed on his damaged cerebellum, which, in healthy brains, makes an important contribution to the early stages of learning.

For most of us, nondeclarative and declarative memory processes are intertwined. You may not be able to describe what you are doing when you ride a bicycle, but you can think back to the days when you rode with training wheels, or when a parent first let go of the back of the bike and let you ride on your own. Skills, experiences, and knowledge are all linked. What remains fascinating about Henry's case is that it showed how a skill could bloom in the brain, even as the experience behind it was irretrievably lost.

Nine

Memory without Remembering II

Classical Conditioning, Perceptual Learning, and Priming

From the mid 1980s through the late 1990s, members of my lab and I expanded our thinking and efforts to investigate the nature of learned behavior. In a broad theoretical context, we designed new experiments to unravel the different cognitive and neural mechanisms that account for nondeclarative memory. As we have seen, Henry was able to unconsciously acquire new motor skills. We also found that he could successfully perform other nondeclarative memory tasks. In our studies of *classical conditioning, perceptual learning,* and *repetition priming,* Henry demonstrated what he had learned through his performance of the tasks and not through conscious declarative memory. His proficiency indicated that these forms of unconscious learning, like motor learning, occur in brain circuits outside the medial temporal lobes. Henry played a major role in the development of thinking about each of these kinds of nondeclarative knowledge.

During this time, my colleagues and I came to realize Henry's limitless worth as a research participant. We continued to be amazed at how many different contemporaneous scientific findings could be related to, or strengthened by, further examining him, and our research with Henry was certainly a boon to my lab's reputation. Although our publications describing his results made up only 22 percent of our total output, these articles were, and continue to be, high profile and widely cited.

Classical conditioning is a learned behavior that capitalizes on a reflex, such as salivation, a kneejerk, or blinking. This form of nondeclarative learning has been for many decades a valuable tool for research in animals and humans. In experiments using classical conditioning, a neutral item, such as the sound of a bell, is paired repeatedly with another item, such as food, which reliably produces a reflex, such as salivation. Eventually, the sound of the bell itself elicits the reflex response. When the subject salivates in response to the bell, we know that during the multiple presentations of the bell with the food, the animal has learned to associate the two.

The Russian physiologist Ivan Pavlov discovered classical conditioning in the early 1900s while studying digestion in dogs. His technique for eliciting this phenomenon capitalized on a simple reflex: when an animal has food in its mouth, it salivates. Pavlov ingeniously observed that a similar reflex could be activated by the smell of the food, by seeing the person who delivered the food, or even by the sound of the person's footsteps. The dogs learned that these sensory cues meant that food was on the way. In Pavlov's experiments, his assistant rang a doorbell just before the dogs received their food. After being exposed repeatedly to these paired stimuli—the bell and the appearance of food—the dogs salivated when they heard the bell, indicating they had learned to associate the sound with food.[1]

Establishing links between items and emotions is a popular strategy for the advertising industry. Picture an ad for a Caribbean resort dominated by beautiful, smiling couples taking sunset strolls on the beach, swimming with tropical fish, and enjoying massages. If we decide to take a tropical vacation, we will likely choose the resort that, thanks to conditioning, we have learned to associate with fun and romance.

My colleagues and I knew from previous experiments that both the cerebellum and hippocampus play a role in the formation of conditioned responses, but we wanted to test how important each area was for this kind of learning. We reasoned that if Henry showed conditioned responses to stimuli without a functioning hippocampus, then his residual cerebellum likely mediated the learning. If Henry did not show conditioned responses, the result would be uninterpretable: we would not know whether the damage to his hippocampus, the cerebellum, or both was responsible for the deficit. All of Henry's neurological examinations dating back to 1962 uncovered signs of cerebellar dysfunction, and his MRI scans showed marked cerebellar atrophy, indicating cell death. But despite his extensive hippocampal and cerebellar damage, Henry did exhibit conditioned responses in our experiments. Although his learning was much slower than that of a healthy man his age, he demonstrated remarkable retention by showing conditioned responses in studies carried out two years after the initial learning sessions.

We first studied Henry's capacity for classical conditioning in 1990, using an eyeblink conditioning experiment, which tested whether he showed a conditioned reaction by blinking in response to a tone that preceded a puff of air in his eye. For testing, Henry sat in a comfortable chair in a quiet room at the MIT Clinical Research Center. He wore a headband that held an air-puff jet and a monitor to record his eyeblinks. The researcher gave Henry these instructions: "Please make yourself comfortable and relax. From time to time, you will hear some tones and feel a mild puff of air in your eye. If you feel like blinking, please do so. Just let your natural reactions take over" (see Fig. 16).[2]

Over an eight-week period, we administered two kinds of conditioning tasks: delay conditioning and trace conditioning. During *delay conditioning*, a tone came first, immediately followed by an air puff, and the two stopped together. Each training session lasted about forty-five minutes and included ninety trials. On eighty of those trials, Henry experienced both the tone and the air puff, giving him the opportunity to unconsciously associate the two. If an eyeblink occurred in the very short interval—less than a second—between the tone and the air puff, we counted it as a conditioned response. This blink indicated that Henry had learned to associate the sound with the forthcoming air puff to his eye, and unconsciously blinked in anticipation of the air puff. On the ten remaining trials, Henry heard the tone alone, and if he blinked immediately, we scored that blink also as a conditioned response. Tallying up the results was simple: we counted how many times he showed a conditioned response in the tone-plus-air-puff condition and in the tone-alone condition. During *trace conditioning*, a silent interval occurred between the tone and the air puff, meaning that Henry's brain had to hold the tone online for half a second to associate it with the air puff that followed. As before, Henry got credit for a conditioned response if he blinked right after the tone.[3]

Throughout the conditioning sessions, we showed Henry films to keep his attention focused on something pleasurable. One of his favorites was *The Gold Rush*, a Charlie Chaplin comedy, and he liked a documentary about the 1939 New York World's Fair, which he had attended with his mother. Although we turned off the sound so Henry could hear the tone, he did not complain and enjoyed the testing experience. All the while, Henry was unaware that he was in a memory experiment, confirming that this task indeed tapped his nondeclarative memory processes. We compared his conditioning scores with those of a healthy sixty-six-year-old man to see whether Henry was impaired and to what extent.[4]

Henry produced conditioned responses in both the delay and trace procedures, an achievement tied to modifications in his brain during the nondeclarative training experience. But overall, his performance was inferior to that of the control participant. He required more trials than the control to reach the *criterion of learning*—blinking in eight out of nine consecutive trials when the tone was presented alone. For delay conditioning, the control attained the goal—eight out of nine correct trials—in 315 trials, whereas Henry required 473. Five weeks after the delay-conditioning experiment, we introduced the trace procedure. For trace conditioning, Henry's control participant reached the criterion of learning on the first trial, whereas Henry required ninety-one trials. It seemed he was impaired on both delay and trace conditioning.[5]

We gained some understanding of Henry's performance during delay and trace conditioning by looking at trials that included the tone but no air puff. On some of these trials, although he produced an eyeblink after the tone, it came too late—after our cutoff for conditioned responses, 400 milliseconds; consequently, these blinks did not count as conditioned responses. His slowness to respond explains, at least in part, why he required over a hundred extra trials to reach the learning criterion for delay conditioning. But a second measure of learning, how much Henry remembered after a five-week break, showed that some of the learning in the delay procedure carried over. This time, Henry required 276 trials to show conditioning—197 fewer than before—while the control participant took ninety-one trials, twenty-four fewer than previously. Although Henry's 42 percent gain was less impressive than the control's 79 percent, he clearly made substantial progress in acquiring the conditioned response. This experiment tells us that the hippocampus is not essential for classical conditioning to occur in either the delay or trace procedure. Henry's ability to learn, although reduced, forces us to speculate about what parts of his remaining cerebellum could have supported this learning.[6]

Two years after the initial conditioning experiment, we examined the durability of this learning. In the new experiments, Henry gave us a striking demonstration of nondeclarative learning: in only nine trials, he reached the criterion of learning for trace conditioning, showing that over the two-year period, the learned conditioned responses were consolidated and securely stored in his brain. This unambiguous result indicated that the hippocampus is not essential for storing a trace of the tone for half a second to associate with the air puff. For conditioning to occur, Henry must have engaged his remaining cerebellum and cortical areas to retain the learned responses for two years. He showed unconscious, nondeclarative learning despite having no declarative memory of the experience: he did not recognize any of the researchers, apparatus, instructions, or procedures, and was unaware of what he had learned.[7]

To better understand the differences between delay and trace conditioning, we turned to the work of three memory researchers at the University of California, San Diego. In 2002, they marshaled evidence from experiments in animals and humans—including patients with amnesia— to highlight that awareness is necessary for trace conditioning but not for delay conditioning. Awareness in this learning task is the declarative knowledge of the relation between the tone and the air puff—the tone signals the imminent arrival of the air puff. Our control participant must have had this declarative knowledge—this awareness—because he acquired trace conditioning in one trial. During the course of the experiment, healthy participants came to understand, on a conscious level, that the tone forecast the occurrence of the air puff, and would come to expect it.[8]

Although Henry lacked this declarative knowledge (awareness), he eventually demonstrated trace conditioning, requiring ninety-one trials. Thus, the mechanism that supported Henry's learning must have been different. The California memory researchers suggested that the cerebellum, while necessary for delay and trace conditioning, could not sustain a representation of the tone across the half-second trace interval. In Henry's case, then, that contribution must have come from a representation of the tone in his intact auditory cortex, which allowed him to acquire the conditioned responses—his nondeclarative knowledge.[9]

Our eyeblink classical-conditioning experiments demonstrated the plasticity in Henry's brain. Through these procedures, he could achieve associative learning, connecting a tone with an air puff to his eye. This nondeclarative learning was involuntary, confined to circuits that operated outside the realm of his conscious awareness. In contrast, if he explicitly tried to associate the tone and the air puff, he would fail, just as he would be unable to associate his doctor's name with his face. He did not have relational, declarative memory circuits to rely on for that task, but he still maintained networks that allowed him to acquire, without conscious recollection, two different kinds of conditioned responses—delay and trace.

Like classical conditioning, perceptual learning is also expressed through performance of a task. *Perception* in the visual system is the mind's ability to detect movement and to identify by sight objects, faces, shapes, textures, orientation of lines, and colors. In the same way, the sense of touch allows the mind to appreciate roughness, temperature, shape, texture, and elasticity.

Perceptual learning is different from perception. It occurs on top of the basic processing of stimuli. Perceptual learning is the ability to identify something more accurately and effortlessly after training, and it occurs incidentally and without conscious awareness of learning. The fine-tuning of perception through experience is apparent in just about any area of life, from the hobbyist who knows every make and model of antique cars, to the quality-control manager of an assembly line who can spot defects in an instant, to the radiologist who can identify a cancerous tumor from the shadings in an MRI image.[10]

My colleagues and I were curious to see whether Henry's medial temporal-lobe lesions would allow him to acquire new perceptual information without being aware that he was doing so. We addressed this question in 1968 when Milner gave Henry a test of perceptual learning, the Gollin Incomplete Pictures Test. This test was not directed at Henry's visual perception, but rather at his ability to identify a less complete version of a picture the second time he saw it compared with the first. The task entailed viewing simple line drawings of twenty common objects and animals, such as an airplane and a duck. Henry viewed each object in five degrees of fragmentation. The test began with a very sketchy representation containing a few parts of each object, which was almost impossible to interpret, and ended with a complete, recognizable picture. Henry first saw the most fragmented set, one drawing at a time, each for about a second, and said what he thought the drawing might represent. He then saw progressively complete sets of pictures until he could name all twenty objects (see Fig. 17).[11]

Milner administered the Gollin test to Henry on two consecutive days, with these instructions: "I'm going to show you some pictures that are incomplete. I want you to tell me what the figure would be if it were completed. Guess if you are not sure." After a short practice test, she showed him the first twenty cards, the most difficult, and noted his errors. She then presented a less-fragmented version of the pictures in a different order so that Henry could not anticipate which drawing would appear next, and told him that this time they would be a little easier to identify. This procedure continued, with the drawings more complete on each trial, until Henry identified the twenty pictures. He completed the test without errors after four trials, and remarkably, his accuracy was a bit better than that of the ten control participants: he misnamed twenty-one pictures, whereas the controls, on average, misnamed twenty-six.[12]

We knew from other testing that Henry's visual perception was excellent, but would this first encounter with the pictures benefit his performance the next time he saw them? Would he show perceptual learning? An hour later and without warning, Milner showed Henry the same set of pictures. Henry did not remember having taken the test before; nonetheless, he identified the fragments in fewer trials.[13]

Still, Henry did not show as much improvement as the controls did. Why not? The control participants had an advantage over Henry: they retained the names of the pictures in their long-term memory, so they had a menu of the correct names to choose from as they viewed the fragmented pictures a second time. They knew, for instance, that one of the pictures was of a duck, so when they saw a few fragments that suggested a beak and tail, they guessed *duck*. But Henry did improve from trial to trial, and amazingly, when I gave him the same test thirteen years later, his identification was even more accurate. Although he still had no conscious memory of ever having seen the pictures, he had learned a perceptual skill without explicit knowledge, and it stuck—solidly stored in the preserved cortical areas of his brain.[14]

We now understand much more about how certain parts of the brain detect and classify information. In the early 1990s, studies revealed, for instance, that one brain area is dedicated to face processing and recognition. Using a functional imaging technique called positron emission tomography (PET), a cognitive neuroscientist at the Montreal Neurological Institute asked participants to identify faces, and found regional increases in cerebral blood flow—indicating heightened neural activity—in areas within the temporal cortex dedicated to the processing of visual information. Five years later, a cognitive neuroscientist at MIT turned to a brain-imaging method that gives more precise pictures of the brain than PET. She developed functional MRI protocols to define the limits of this face-selective area in the temporal lobe and establish its function, christening it the *fusiform face area*. This area was not damaged in Henry's brain, so he could still recognize his parents, relatives, friends, and celebrities after his operation; he had stored those images in his long-term memory preoperatively. If we had shown Henry a series of unfamiliar faces inside an MRI scanner, his fusiform face area would have been active while he was looking at them. But after he left the scanner, he would not remember them, because he lacked the necessary areas in his medial temporal lobe to form these new memories.[15]

The MIT researcher's pivotal discovery inspired a team of scientists at Vanderbilt University to conduct further functional MRI studies, showing how the brain maps other kinds of expertise. They found that extensive knowledge of birds or cars also recruited the face-selective area in the brain. In the MRI scanner, all participants saw pairs of cars and pairs of birds, and judged whether the cars were the same model but different years, and whether the birds belonged to the same species. When the researchers compared brain activity associated with cars and birds in the two groups, she found greater activation for cars than birds among the car aficionados, and greater activation for birds than cars among the bird experts. The car- and bird-expertise effect occurred in the same cortical area as face expertise, suggesting that activity within this small area is differentially focused to support several specializations—face recognition and expert recognition of objects.[16]

These experiments illustrated plasticity within individual human brains—the dedication of neurons in a precisely defined area resulting from long-term perceptual learning with specific objects, such as faces, cars, and birds. This ability is fundamental to successful interactions with other people and our environment. After his operation, Henry could still perceive faces, cars, and birds, and could show normal perceptual learning with the Gollin pictures—capacities that relied on his intact visual cortex. But those processes, by themselves, were insufficient for him to remember, in the everyday sense of the word, new faces and objects.

We have continued to learn more about how our brains learn and classify information. In 2009, neuroscientists identified white-matter pathways connecting the visual areas that support face and object processing with the amygdala and hippocampus. We will examine Henry's autopsied brain to confirm the integrity of these connections. We assume, however, that these pathways to the medial temporal-lobe structures were intact, so information about faces and objects would have reached his medial temporal-lobe structures if they had not been removed. Henry lacked the machinery to receive, encode, and consolidate face information as a memory.[17]

Not every kind of nondeclarative learning requires repeated exposures to a stimulus or procedure. *Repetition priming* can occur after one learning trial. In the lab, when Henry looked at a series of words, pictures, or patterns and encountered them a second time in a subsequent test, his perception of them or his response was often facilitated due to his earlier exposure. This enhanced processing is referred to as repetition priming—his brain was "primed" to respond in a certain way to a stimulus because he had seen it previously. Even when he was not intentionally trying to recall the past, his experience unconsciously influenced his memory.[18]

Repetition priming occurs frequently in our daily lives but usually goes unnoticed. We may hear a song on the radio first thing in the morning and then find ourselves humming it during the day without knowing why. Priming is a favorite tool of the advertising industry. Frequent exposure to particular brand names on TV or in magazines may prime us to process them more and thus select them above other brands even if we do not consciously remember seeing them in advertisements. Political campaigns also take advantage of priming: little-known candidates can become popular overnight if constituents see and hear their names repeatedly. When we read these names on our voting ballot, we may mistakenly think that the candidates are seasoned politicians with formidable track records, simply because we process their names more easily.

In the mid 1980s, we became interested in examining in detail Henry's ability to prime because we wanted to know whether this form of memory was resistant to amnesia and whether different kinds of priming were equally robust in amnesia. Another focus of our research was demonstrating whether the priming effect was comparable when the test items were familiar to Henry, compared to those that were novel.

We explored these topics in a series of experiments carried out in the late 1980s and throughout the 1990s, using a variety of priming tasks. In every case, the test had two parts: a *study phase*, in which Henry was exposed to words or pictures, followed by a *test phase*, in which he performed a task with studied and unstudied words or pictures. For example, in one study phase, we showed him a list of words on a computer screen, one at a time, and asked him to say "Yes" if the word had the letter *A* in it and "No" if it did not. This instruction led Henry to believe that we were simply testing his ability to detect *As*, so he did not think that this was a memory test.

EPISODE
FACULTY
RADIUS
STOVE
CALCIUM
ROUGH
CLAY
STAMP
FROST

Next, in the test phase, we showed Henry the first three letters of these words interspersed with the first three letters of comparable words that were not in the study list.

CLA
SER
CAL
ROU
MED
TRO
EPI

FAC
SWI
RAD
BRE
REC

We told Henry that each three-letter stem was the beginning of an English word, and asked him to make each stem into a word. We encouraged him to write the very first word that popped into his mind, and did not mention the study list. Henry remained unaware that his memory was being tested.

The studied words were not the most common completions of the stems in that they were not among the three most popular responses given in a pilot study in which we asked healthy participants simply to complete stems with the first word that came to mind. Common completions of the stems CLA, CAL, and ROU included *clap*, *calendar*, and *round*. *CLAY*, *CALCIUM*, and *ROUGH* were more unusual choices. Remarkably, after a single exposure to the study list, Henry gave the less-common completions, indicating a priming effect. His priming score took into account the number of items he would get correct just by chance. It was the number of times he completed the stems to the studied words minus the number of times he completed other stems to comparable unstudied words—words similar to the studied words in number of letters and frequency of occurrence in the English language. During the test, priming occurred as the result of activation in Henry's brain of the representation of the word he had just encountered in the study list.[19]

We compared Henry's performance on this nondeclarative memory task with his scores on two measures of declarative memory, in which his task was to consciously recollect studied words in a similar format. He viewed the study list on a computer screen as before, and after a short delay, we asked him to orally recall the words he had just seen. Next, we gave him a recognition-memory test, in which three words appeared on the computer screen, all of which began with the same three-letter stem, for example, *CLAY—CLAM—CLAP*. Henry's task was to select the studied word from among the three choices. He was impaired on both measures—recall and recognition.[20]

The critical difference between these declarative tasks and the nondeclarative priming task was in the instructions. For the recall and recognition tests, we asked Henry to intentionally retrieve words from the study list—a memory test in the traditional sense. These results showed that he activated separate neural networks for declarative and nondeclarative learning. He failed the recall and recognition tests, showing that his declarative circuits were faulty, but performed normally on the word-completion priming test, proving that his nondeclarative circuits were preserved.[21]

What brain mechanism allows people with amnesia to show normal priming? The first hints at an explanation came in 1984, when psychologists at the University of Pennsylvania made an astute observation during casual conversations with amnesic patients. The researchers noticed that densely amnesic patients, after lengthy exposure to a particular word or concept—for example, dogs or types of dogs—followed by fifteen seconds performing another task, would claim to have no memory for any particular conversation, and no idea what the topic of conversation might have been. But if the researchers then asked the patients to initiate a conversation on any topic they wished, they were likely to choose a topic or mention a word from the previous discussion—*dogs* or *terrier*, for example—even though they did not recognize the link between the new discourse and the prior conversation.[22]

The researchers speculated that amnesic patients' normal performance on priming tests resulted from *trace activation*, the excitation of intact mental representations—symbolic codes for information. They proposed that when participants read words out loud, such as *CANDLE*, *PLEASANT*, and *BUTTON*, they activate a mental depiction of that word. This activation lasts for minutes or hours—just as your hair dryer stays hot for a short time after you have switched it off—and it occurs similarly in normal and amnesic participants. On a subsequent test, when participants have to complete *CAN*, *PLE*, and *BUT* to the first word that comes to mind, the words *CANDLE, PLEASANT,* and *BUTTON* are highly activated and, therefore, more likely to be selected than are other possible completions.[23]

In the mid 1980s, when my colleagues and I began to study the repetition priming-effect, we had several goals. One was to examine nonverbal priming by using unfamiliar patterns as test stimuli. Most demonstrations of intact priming in amnesia had used verbal tasks, such as reading, spelling, or word completion, but a larger theory of the nature of priming in amnesia would have to encompass information other than words. When amnesic patients see words, they can draw on the knowledge about those words that they acquired before they became amnesic: the stimuli are already stored in their mental dictionary, and can be activated and thus primed. But what about information they are seeing for the first time? It was possible that amnesic patients would show intact priming only when they had knowledge of the primed response—when they already possessed a normal representation of the stimulus. Researchers could easily identify knowledge of words as the basis of verbal priming, but it was less clear what constituted a knowledge base for nonverbal (pattern) priming.

In 1990, members of my lab set out to discover whether Henry would show normal priming when the stimuli were patterns drawn on paper. We created six target figures by connecting five dots from the nine possible dots in a three-by-three square matrix. We then asked Henry, along with a group of control participants, to draw any figure they wished, using straight lines to connect the five dots in each of the six dot patterns. These figures constituted the participants' baseline figures, indicating what figures they would choose to draw spontaneously on their own. The priming test came six hours later. In the *study phase*, participants received a sheet of paper with the six target figures and were told to copy these figures onto corresponding dot patterns on the same page. We then removed this sheet of paper, and participants performed a distractor task for three minutes—writing down the names of as many famous entertainers from the twentieth century as possible (see Fig. 18).[24]

In the *test phase*, we gave Henry and the control participants a new sheet of paper showing the six dot patterns, and asked them to draw any figure they wished, as long as they connected the five dots in each pattern with straight lines. We wanted to see whether participants drew the target figures they had copied earlier; if they did, it was proof they had been primed. The number of target figures Henry and the controls drew in the primed (post-copying) condition far exceeded the number of target figures they had produced by chance in the baseline condition. In short, after participants copied a target figure onto a dot pattern, they were more likely to draw that target figure when asked to draw whatever they wanted. Henry showed a normal magnitude of priming on three different test forms administered on three separate occasions.[25]

This demonstration of priming with novel stimuli suggested that the learning was tied, not to Henry's preoperatively established memory representations, but instead to newly acquired representations of the specific target figures. This discovery was the first report of intact nonverbal priming in a memory-impaired individual, providing strong evidence that the priming spared in amnesia is not limited to language-based stimuli.[26]

How do we account for pattern priming? It seemed unlikely that the normal participants or Henry had preexisting memory representations of the target figures, so it is difficult to describe pattern priming as the activation of long-term memory representations. What, then, is the alternative explanation? In the course of copying the target figures onto the dot patterns, Henry and the controls formed new associations between them. The new associations influenced perceptual processing that assigned a specific structure to the dot pattern and guided the primed drawings. Henry's severe amnesia eliminated the possibility that his intact pattern priming reflected the operation of recall and recognition mechanisms, underscoring the conclusion that new associations supporting perceptual priming may be established nondeclaratively, despite severe deficits in episodic memory.[27]

Notably, on a pattern-recognition task, when declarative memory *was* required, Henry's performance was significantly weaker than the controls'. We administered another test in which we instructed Henry and the controls to copy a new set of target figures onto dot patterns, and after a three-minute break, to select from four figures the one they had just copied. Henry, consistent with his poor declarative memory, had trouble recognizing the target figures he had just copied, whereas the control participants did not.[28]

Henry's capacity for pattern priming demonstrated that this kind of memory did not rely on the medial temporal-lobe structures that supported recall and recognition memory. Instead, the perceptual associations that mediate pattern priming are likely established in the early stages of visual processing, located in the back of the cortex. These associations are relatively inaccessible to conscious awareness. This observation gave rise to further experiments—a broad quest for the specific cortical circuits that support the various kinds of priming. We conducted a series of studies that uncovered the functional architecture of repetition priming. Henry played a major role in this research, but we also needed participants whose brains had damage in other areas. We, therefore, recruited patients with Alzheimer disease and others with lesions in discrete brain areas. In addition, we tested a control group of healthy adults who were comparable to each patient group in terms of age, sex, and education.

Our first breakthrough came in 1991 when we demonstrated that priming is a multipart concept: priming represents a family of learning processes. By studying patients with Alzheimer disease, we were able to show that separate circuits in the cortex mediate two different kinds of priming. Like Henry, Alzheimer patients have damage to medial temporal-lobe structures and are impaired on measures of declarative memory, such as recall and recognition. They also have cell loss in certain cortical areas but not others. Our experiment revealed that Alzheimer patients had normal priming when we asked them in the test phase to visually identify words—*perceptual identification priming*—but not when we asked them to generate words based on their meaning—*conceptual priming*. This finding indicated a clear distinction between these two types of priming, suggesting that priming based on simple visual memory depends on a different brain network from priming based on more complex thinking. Henry showed normal priming on both measures because they did not require the participation of medial temporal-lobe circuits.[29]

The perceptual and conceptual priming tasks consisted of a study condition and a test condition. The study condition was the same for both measures—patients and control participants saw a series of words one at a time on a computer screen and read each word aloud. The test conditions differed. For *perceptual identification priming*, the experimenter told participants that they would perform another task unrelated to the one just completed. She then presented a series of words briefly on the screen and instructed participants to read each word. Half the words had been in the study list, and half were new words. Priming was present if the time—measured in milliseconds—needed to identify the studied words was less than the time needed to identify the unstudied words. We discovered that the priming effect in the Alzheimer group did not differ from that in the matched control group, so mild-to-severe dementia did not interfere with priming in perceptual identification. This finding indicated that the circuits in the Alzheimer brains that supported this kind of priming were undamaged.[30]

In the test for *conceptual priming*, participants saw three-letter stems on the computer screen and completed each one with the first word that came to mind. Half were words they had seen before in the study list, and half were new words. This time, when the Alzheimer group had to make each stem into a word, they completed no more of the stems to the studied words than they would have by chance. Their magnitude of conceptual priming was significantly depressed.[31]

From autopsies of Alzheimer patients, we knew that the disease does not damage the cortex in uniform ways. The cortical areas that receive basic information through vision, hearing, and touch, as well as the cortical areas that issue motor commands, are relatively spared, but the high-order areas in the frontal, temporal, and parietal lobes that support complex cognitive processes are compromised. Our priming study implied that a memory network within visual areas in the occipital cortex—intact in Alzheimer disease—supported perceptual priming effects, whereas a different network in the temporal and parietal cortices—damaged in Alzheimer disease—supported conceptual priming effects.

All of these areas were intact in Henry's brain, which is why he had no problem with both kinds of priming.[32]

In 1995, our investigation of a patient with damage to the visual areas in his brain buttressed the argument that perceptual and conceptual priming processes are separate. This man's MRI scans showed multiple areas of abnormality, particularly in his visual areas, and he had marked deficits on tests of visual perception. He did not have amnesia, however, and his medial temporal-lobe structures were spared. We gave him the same tests the Alzheimer patients had performed, and the results showed the reverse pattern. He had no capacity for perceptual priming, identifying briefly presented words and pseudowords, but he did show normal conceptual priming, completing words based on meaning, and could still explicitly recognize words he had seen before. The dramatic contrast between this man's normal performance on the conceptual priming task and his lack of priming on perceptual priming tasks was a vital addition to our thinking. When considered beside the opposite distinction observed in the Alzheimer patients, the results provided convincing proof of the existence of two priming processes that depend on separate neural circuits.[33]

Our studies of repetition priming with Henry and other patients helped reveal in increasingly finer detail how our experiences influence us without our explicit knowledge. Because we included measures of declarative memory in our priming experiments, the results highlighted the distinction between priming—nondeclarative memory—and explicit retrieval—declarative memory. This dissociation that we teased apart so meticulously in the lab is also blatantly apparent in daily life. When we forget appointments or friends' birthdays, our memory has failed us; but if we lose a tennis game, we do not blame our memory and say, "I couldn't retrieve the correct motor sequence for my serve." By using the term *memory* in the first case but not the second, we acknowledge that declarative and procedural memories are different.

But anecdotes from daily life do not prove that such a distinction exists in the functional organization of the brain. We needed Henry and other patients to make that dissociation stick scientifically. Henry had already shown us that the hippocampus and neighboring tissue were critical for declarative memory—the ability to intentionally remember experiences and information. His normal performance in our priming experiments supports the view that conceptual and perceptual priming are localized to memory circuits embedded in high-order association cortex in the frontal, temporal, and parietal lobes—areas known to support complex cognitive functions. These circuits work independently from the medial temporal-lobe memory circuits.

Henry was important to our understanding of the various kinds of memory that operate outside of conscious awareness. Our studies of classical eyeblink conditioning, perceptual learning, and repetition priming further uncovered his capacity to acquire new nondeclarative knowledge. Despite massive damage to his hippocampus and surrounding structures, on both sides of his brain, resulting in profound amnesia, he could learn, without using explicit retrieval processes and without consciously recollecting the learning episodes. He acquired conditioned responses during delay and trace conditioning and retained these learned responses months later; he mentally completed picture fragments, showing the benefit of prior exposure to those pictures; and he primed with both verbal and pictorial test items. These accomplishments testify to Henry's residual cognitive abilities and the neural circuits that support them.

Members of my lab and I eagerly communicated the results of our experiments on nondeclarative learning and memory to the medical and scientific communities in the form of articles in scientific journals and book chapters. Recognition of Henry's contributions is evident in the hundreds of citations of our work by other researchers.

Ten

Henry's Universe

Henry's mother continued to take care of him for several years after his father's death, but eventually the responsibility became too much for her. In 1974, when Henry was forty-eight, he and his mother moved in with Lillian Herrick, whose first husband was related to Henry on his mother's side of the family. Mrs. Herrick was a registered nurse who before retiring had been employed at the Institute of Living, an expensive psychiatric treatment facility in Hartford, Connecticut. In her sixties, she sometimes took in older people who needed help in their daily lives.

Mrs. Herrick and her husband lived in an established residential neighborhood on New Britain Avenue in Hartford, near Trinity College. Their large three-story white wooden house had a front porch and was surrounded by tall trees. Mrs. Herrick's son, Mr. M., described her as "prim, proper, and very English." She had a good sense of humor and laughed a lot. At home, she wore old-fashioned housedresses but also enjoyed dressing up to go out. Her son never saw her in a pair of pants.

Even though Mrs. Herrick's first husband had died, she maintained her connection to the Molaisons, taking pity on Henry and keeping in touch with him and his mother for years. This bond was fortunate for Mrs. Molaison, who was aging and becoming infirm. On one visit, Mrs. Herrick was shocked to find that Mrs. Molaison had a large, terribly inflamed ulcer on her right leg. Mrs. Herrick immediately drove her to the ER at Hartford Hospital, and for the next two days she was in danger of losing her leg. Fortunately, her leg healed, and after that incident Mrs. Herrick checked in on Henry and his mother every two or three weeks.

In December 1974, Mrs. Herrick received a telephone call from Molaison family friends who lived in the neighborhood, reporting that when they took a Christmas package over to her house, Mrs. Molaison did not recognize them. Mrs. Herrick was scheduled to go on duty at the Institute of Living but called in to say that she was unable to work that day, and instead drove to the Molaison home. She described Mrs. Molaison as "on the floor and completely out of it." It is unclear what happened to her, but Henry seemed unaware that anything was amiss; he thought that his mother was just resting or sleeping. An ambulance transported Mrs. Molaison, in a very bad state, to the ER. The doctors wanted to send her directly into a nursing home, but in January 1975, kindhearted Mrs. Herrick accepted Mrs. Molaison and Henry into her own home.

Mrs. Herrick immediately noticed that their personal hygiene was deplorable, including soiled underwear and excessive body odor. She improved their personal care and, in her words, got Mrs. Molaison "back to where she was very good for a long time." At Mrs. Herrick's house, Henry's relationship with his mother was initially stormy. They may have had conflicts in the past, but no one had had a chance to observe them up close. According to Mrs. Herrick, Mrs. Molaison nagged her son constantly, and he would become "really, really angry" with her, kicking her in the shin or hitting her on the forehead with his glasses. Mrs. Herrick soon intervened and relegated Mrs. Molaison to the upstairs of her house and Henry to the downstairs. If they were together, Mrs. Herrick stayed in the room with them to keep the peace. This strategy worked, and Henry settled down considerably.

Mrs. Herrick introduced a routine into Henry's life. Every morning, he would have his breakfast, take his medicine, shave, and go to the bathroom. She would remind him to get clean underwear and socks from his drawer and get dressed. On weekdays, at quarter to nine, Mr. or Mrs. Herrick drove him to a "school" for people with intellectual disabilities—HARC, the Hartford Association for Retarded Citizens. Henry and a few others sat around a table doing piecework submitted by Hartford businesses, such as placing key chains on a cardboard display. In return, they received a small check every other week.

In June 1977, Henry's vocational progress report noted that he "has adjusted well to the workshop." His instructor wrote this description of Henry's "work assets":

> Henry does not retain instructions well. Periodically, must be re-instructed. Is willing to adapt to job changes but gets confused. Perseveres at his work task. Henry's work must be checked occasionally. Henry's work does not improve with repetition. The quality of his work decreases as the number of steps to the task increases. Has difficulty with multi-step assembly tasks. Can follow verbal instructions.

The instructor specifically noted that Henry could not handle a project that had more than three steps to it.

After his work breaks, Henry often went to the office to ask what he was supposed to do, but as soon as he was shown his desk, he knew exactly what his task was. The context helped him remember the procedures that comprised his work, skills he had stored away in his non-declarative memory circuits, which could be activated in response to the appropriate environmental cues.

Back at Mrs. Herrick's house after school, Henry's routine included washing his hands and having a snack. He liked to sit on the patio with his rifle magazines and crosswords, and if others were outside, he would talk with them. He was much more sociable in that setting than he had been when living alone with his mother. Henry wanted to be useful at home; he would take out the trash cans and help Mr. Herrick with yard work. In the evening, he sat in an overstuffed armchair, watching television or doing crossword puzzles. Mrs. Herrick posted a sign on the television set saying that it was to be shut off at nine-thirty, and Henry always obeyed; he willingly went to bed at nine-thirty or ten. Having been raised Roman Catholic, Henry watched one or more masses on TV on Sunday morning, and afterward Mrs. Herrick would often take him out for a drive and dinner. He loved to go out to dinner. These afternoon outings lasted several hours. Henry was not fussy about their destination, just delighted to go anyplace she took him. "He goes as long as the car will go," she said.

Henry did not get lost in Mrs. Herrick's house. He knew where his room was and was conscientious about turning off his lights. He was cognizant of safety around the house. On one occasion, Mrs. Herrick had something cooking on the stove, and Henry, thinking she had gone off and left it, turned off the gas. One night, she went upstairs to set her hair and told Henry to leave the light on in the kitchen because she would be down later. After she left, however, Henry had trouble remembering with certainty what her instructions had been, so instead of going to bed he waited downstairs forty-five minutes until Mrs. Herrick returned.

My conversations and correspondence with Mrs. Herrick assured me that she was taking excellent care of Henry and creating a warm but disciplined environment for him. When he moved into her house he was a heavy smoker with a daily pack-and-a-half habit; Mrs. Herrick gradually cut him down to about ten cigarettes a day, and eventually to five. At some point during the six years Henry lived with Mrs. Herrick, his chest X-ray during a physical exam showed emphysema, so she cut him off completely. After he stopped smoking, his complaints of stomach pains subsided, but I suspect that the urge to smoke lingered. Around this time, while I was testing him, he automatically reached into his chest pocket; when I asked him what he was looking for, he replied, "My smokes." His old habit was tenacious. Henry's nondeclarative memory was intact—he could remember the gesture of reaching for his cigarettes, which he had learned before his surgery. His declarative memory, however, was gone—he could not recall why his pocket was empty.

To ensure proper hygiene, Mrs. Herrick left notes around the house reminding Henry to do things such as wash his hands and raise the toilet seat. He seemed to be in better health, more alert, and eating a more varied diet than when he had lived alone with his mother. Henry stuck to his routine; he missed schooldays only when he had a major seizure and was lethargic afterward. These grand mal seizures were infrequent, but he still had quite a few petit mal episodes—temporary absences. According to Mrs. Herrick, he would be watching television and suddenly "just go blank," returning to his usual self within seconds. She looked after his medical care and coordinated his visits to our lab, graciously driving him to MIT for testing whenever we wanted to see him.

Mrs. Molaison also benefited from Mrs. Herrick's attention. In February 1977, however, she had what Mrs. Herrick called "another bad spell" and was hospitalized with high blood pressure. She left the hospital after a week, but at age eighty-nine, she clearly required more care than Mrs. Herrick could give. Mrs. Molaison went to live in a nursing home, where she spent the rest of her life, demented and delusional. Without the ability to recollect where his mother was and why, Henry had trouble adjusting to her absence. He often asked when his mother and father were coming to visit him. That year, one of our lab members noticed that he had written two notes to himself, which he kept in his wallet, one saying "Dad's gone," and the other "Mom's in nursing home—is in good health." We do not know whether Mrs. Herrick prompted him to write these notes or whether he did it on his own initiative when she gave him this information, but either way the notes protected him from the anxiety of not knowing where his parents were.

Mrs. Herrick occasionally took Henry to visit his mother. He was always happy to see her and just as happy to leave, reassured that she was OK. She died in December 1981 at the age of ninety-six. According to one of Henry's caregivers, he did not take the news of her death too badly and was not overwhelmed with grief. He spoke only of what a nice woman she was and described how she had taken care of him all his life.

Henry continued to live with Mrs. Herrick until 1980, when she was diagnosed with terminal cancer. Henry, in his mid-fifties, then moved to nearby Windsor Locks, Connecticut, to Bickford Health Care Center, a long-term care facility founded by Mrs. Herrick's brother, Ken Bickford, and his wife Rose. Bickford Health Care was a friendly environment where Henry received round-the-clock attention from a large staff of specialized, dedicated caregivers for the remaining twenty-eight years of his life. His hospital chart initially listed me as "the only interested relative, friend, or contact." I was there on the day of his admission to see that he would be well cared for and protected. Medicare, Medicaid, and Social Security covered the cost of Henry's stay at Bickford, as well as his visits to local hospitals.

With Mrs. Herrick gone, I became Henry's sole keeper, in the sense that I felt responsible for his well-being. I watched over him. When he visited us at the MIT Clinical Research Center, he always had a physical and neurological exam, which helped us pinpoint any new symptoms and find ways to alleviate them. Luckily, we could rely on the resources of the MIT Medical Department and the nursing home staff to carry out the physicians' orders. I kept in close touch with Henry's Bickford caregivers, who always called me when a new concern arose, such as when he had a grand mal seizure, broke his ankle, or showed obstreperous behavior. I also tried to enhance his quality of life by sending him clothes, cards, photographs, and movies and the equipment to play them.

I was now in the position of knowing more about Henry than any living person. Mrs. Herrick had become the custodian of the Molaisons' mementos, collected on vacations and from family events, and she passed them on to me. In 1991, the Probate Court in Windsor Locks, Connecticut appointed her son, Mr. M., to be Henry's conservator, meaning that he was responsible for protecting Henry's interests and supervising his personal affairs. He was my best resource for information about Henry's past, filling me in on details of the Molaison family history and giving me a treasure chest of mementos, which on several occasions I have had the pleasure of sharing with the general public. All these stories and keepsakes have helped me reconstruct the family's past.

We might see the five decades of Henry's life with amnesia—first with his parents, then with Mrs. Herrick, and finally at Bickford—as extremely impoverished. Although he was always looked after, could amuse himself, and rarely appeared to suffer, what kind of life could he have lived without his memory? If he were forever trapped in a single moment, could he be a fully realized human being? Some philosophers, psychologists, and neuroscientists have argued that without memory, we lack identity. Did Henry have a sense of who he was?

There is no doubt in my mind that Henry did have a sense of self, even though it was fragmented. Over the years of working with him, we came to know his personality and the quirks and traits that made him who he was. Henry's beliefs, desires, and values were always present. He showed a general spirit of altruism and frequently articulated his hope that what we learned about him would help others. That possibility was a source of satisfaction for him.

Henry knew that he had undergone an operation and was aware that he had trouble remembering things, but he had no idea how far back in time his memory loss extended. Here is how he talked about his operation in a 1992 conversation with me:

SC: Tell me about that [the operation].

Henry: And I remember it if—I don't remember just where it was done in—

SC: Do you remember your doctor's name?

Henry: No, I don't.

SC: Does the name Dr. Scoville sound familiar?

Henry: Yes, that does.

SC: Tell me about Dr. Scoville.

Henry: Well—well, he went—he did some traveling around. He did—well, medical research on people. All kinds of people in Europe, too, and the royalty, and out in the movie stars, too.

SC: Did you ever meet him?

Henry: Yes, I think I did. Several times.

SC: Do you know where you met him?

Henry: I think I met him in his office.

SC: And where was that?

Henry: Well, I think of Hartford right away.

SC: Where in Hartford?

Henry: Well, tell you the truth. I can't tell you the address number or anything, but I know it was down the main part of Hartford in—but it was off the main section. Off of Main . . .

SC: Was it at a hospital?

Henry: No. The first time that I met him was in his office. Before I went to a hospital. And there—well—well, what he learned about me helped others too, and I'm glad about that.

Henry's recollections were largely correct. While he never expressed any resentment toward Scoville or about the outcome of the operation, he did seem to have processed on some level that something very bad had happened as a result of his surgery. Henry mentioned many times that he had once dreamed of becoming a brain surgeon, but said he had given up on the idea because he wore glasses and was worried that he might make a mistake and hurt a patient. It was not unusual for him to repeat versions of this little story three or four times a day. In one scenario, an attendant wiping Henry's brow inadvertently dislodged his glasses; in another, blood spurted up on his glasses and obstructed his vision; and in a third, little specks of dirt on his glasses blocked his view. In all versions, Henry expressed concern that he would make a wrong move, resulting in the patient's sensory loss, paralysis, or death. The resemblance between these recurring narratives and Henry's description of his own experience is striking. In 1985, Henry shared these thoughts with a postdoctoral fellow in my lab, Jenni Ogden, a neuropsychologist from New Zealand:[1]

Ogden: Do you remember when you had your operation?

Henry: No, I don't.

Ogden: What do you think happened there?

Henry: Well I think it was, ah—well, I'm having an argument with myself right away. I'm the third or fourth person who had it, and I think that they, well, possibly didn't make the right movement at the right time, themselves then. But they learned something.

Henry's kindness was apparent in many routine ways. Socially, he was courteous, friendly, and chivalrous. When we walked together from one MIT building to another, he would cup my elbow with his hand to escort me down the sidewalk. He also had a sense of humor and enjoyed cracking jokes, even at his own expense. In 1975, during a conversation with one of my colleagues, Henry gave his usual line, "I'm having an argument with myself," in response to a question about the date. My colleague joked, "Who is winning the argument? You or you?" Henry laughed, repeating, "You or you." Henry snapped at me only once in forty-six years: I was trying to help him learn a complex procedure, and he had become frustrated. "Now you've got me all balled up!" he scolded me.

In shaping the Henry we knew, several variables were likely at play: his inherent nature, his protected living situations, and his operation. His behavior was influenced in part by the removal of his left and right amygdalae. A component of the limbic system, this almond-shaped structure is critical for processing emotion, motivation, sexuality, and pain responses, particularly feelings of aggression and fear. Was this sweet, tractable man pacified by his operation? From what we know of Henry, he had always been an agreeable, passive person—similar in behavior to his father—and his parents made no mention of any personality change after the operation. Indeed, Henry had not lost his capacity for emotion. He could even be aggressive—when he attacked a staff member at the Hartford Regional Center, or when he fought with his mother. He also was capable of feeling grief for lost loved ones. As we saw during Henry's stay at MIT in 1970, he was able to miss his mother and to show tenderness toward her when he saw her after an extended absence. Henry's emotions may have been blunted by his operation, but he was able to take part in most of the feelings we all experience.

Still, Henry lacked self-awareness in many basic ways. He was largely unable even to evaluate his own physical state—whether he was sick or well, energetic or tired, hungry or thirsty. Henry's complaints of physical pain were infrequent. He sometimes reported ailments such as stomachaches and toothaches, but other conditions, such as bouts of hemorrhoids, went without comment. When he broke his ankle, he considered the injury so trivial as to not warrant an X-ray. We also noticed that Henry rarely spoke of being hungry or thirsty, but when asked if he was hungry, he would say, "I can always eat." In 1968, Mrs. Molaison reported that, for the first time, Henry agreed with her when she told him that he should be hungry. He said, "Yes, I guess I'm hungry." He never sought out food for himself; it was simply given to him by his caregivers.

How much of Henry's apparent inability to register his internal states was a result of his amnesia, and how much of it was due to the missing amygdalae? To systematically document our observations that Henry rarely commented on such internal states as pain, hunger, and thirst, we conducted two experiments in the early 1980s. In one study, we tested his ability to perceive pain, and in another, we asked him to rate his feelings of hunger and thirst before and after meals. Healthy control participants also performed both tasks. Because limited memory ability might have influenced Henry's reporting of internal states, we compared his performance with that of five other amnesic patients whose amygdalae were spared.[2]

Neuroscientists have been studying the amygdala since the early nineteenth century, with an evolving understanding that this structurally and functionally diverse unit plays a role in a host of behaviors, including pain, hunger, and thirst. Because Henry had almost all of his left and right amygdalae removed, it was important to document the effects of these lesions on its known functions. Each amygdala is part of a dedicated pain-processing circuit that incorporates two other areas— one in the midbrain, the periaqueductal gray matter, and the other just under the frontal lobes, the anterior cingulate cortex. This network evolved to protect animals and humans from adversity and to enhance their chances for survival. The amygdala, in concert with several other brain regions, including the hypothalamus, also contributes to the appreciation of hunger and thirst.[3]

In 1984, my lab began to examine the extent to which Henry could process signals related to pain, hunger, and thirst by first testing Henry's pain perception, using a hairdryer-like contraption that projected a spot of heat onto his skin. We instructed him to apply heat at different levels of intensity to six places on his forearm. The heat was never hot enough to burn his skin. During three test sessions, he rated the intensity of each heat stimulus on an eleven-point scale—*absolutely nothing, maybe something, faintly warm, warm, hot, very hot, very faint pain, faint pain, pain, very painful,* and *withdrawal (intolerable).* We evaluated Henry's pain perception on three occasions. Our analysis of his responses yielded two measures of pain perception—how well he could discriminate between two stimuli of different intensities, and his propensity for calling a stimulus painful. When we compared Henry's performance to that of the healthy control group, he was impaired on both measures. Not only did he have more difficulty than normal participants in discriminating between levels of heat, meaning that he tended to confuse the stimuli, he also did not label any of the stimuli as painful, no matter how intense they were. Remarkably, he never withdrew the heat before the three-second interval ended. The performance of the other amnesic patients was similar to the controls', indicating that Henry's deficit in appreciating pain was not an obligatory component of amnesia. Instead, his amygdala lesions caused this deficit in pain perception.[4]

In another experiment, we compared Henry's ability to perceive the intensity of his hunger to that of healthy participants as well as other amnesic patients. At mealtime, most of us can mentally look inward and access our hunger—do we or do we not want to eat? Then, after finishing a meal, we have a conscious sense of fullness in our stomach, a feeling that tells us whether to skip dessert. When we investigated whether Henry would experience these two measures of appetite, we found that his subjective appetite (*How hungry am I?*) and his sense of fullness (*How full am I?*) were both deficient.[5]

In 1981, we asked Henry to rate his hunger on a scale from zero (famished) to one hundred (too full to eat another bite) both before and after his meals. He consistently gave a rating of fifty whether he was about to eat or had just finished. One evening, after he had eaten a full dinner and his tray was whisked away, a kitchen staff member replaced it, without comment, with another tray containing a meal identical to the one he had just finished. Henry ate the second dinner at his usual slow but steady pace, until only the salad remained. When we asked him why he had not eaten his salad, he simply said that he was "finished," not that he was completely full from having eaten too much. Twenty minutes later, we asked him to rate his hunger again. He gave a score of seventy-five, meaning that he was conscious of being somewhat full. Only by making doubly sure he was full had we finally brought him above the rating of fifty, but he still fell far short of reporting that he was satiated.[6]

The pain-perception test indicated that Henry's capacity for pain detection was disproportionately compromised, compared to his capacity to detect light touches to his skin, which was normal. Although he could discriminate among different levels of pain intensity, his scores were inferior to those of other amnesic patients and healthy controls. His reports of pain did not increase as the intensity of the heat increased.

Because abnormal pain perception was not observed in the amnesic patients whose damage left the amygdala intact, we reasoned that Henry's abnormal tolerance of pain was caused by the removal of both amygdalae. The related findings that he showed no difference in his ratings of hunger or thirst from before and after a meal, and that he was unable to express a feeling of satiety, supported our conclusion that information about current internal states was either lacking or less accessible to Henry than to other amnesic patients. We attributed his failure to label and express his internal states—pain, hunger, and thirst—not to his memory deficit, but instead to his amygdala lesions.

Our experiments confirmed what we had seen in Henry's daily life: a failure to appreciate pain and poor monitoring of his appetite. We concluded that the bilateral resection of his amygdala accounted for his poor appreciation of internal states. Without his amygdalae, Henry did not sense when he was hungry or thirsty, and could not engage the brain circuits that told him he had had enough to eat and drink. Fortunately, his overall appreciation of food was not diminished. He told us that he preferred cake to salad, was very fond of French toast, and disliked liver.

The amygdala also plays a role in the expression of sexual drive, and lesions of the amygdala may increase or decrease a patient's libido. To the best of our knowledge, Henry did not show any sexual interest or behaviors after his operation. In 1968, fifteen years postoperatively, Scoville wrote that Henry "has had no sexual outlets, nor does he appear to have a need of them." Henry's lack of libido may have been a casualty of his operation. He had mentioned interactions with girls in his youth, and letters he received from two friends suggested that he had been interested in women prior to his operation, although he apparently did not have any serious romances. A photograph in Henry's family album of an attractive young woman in a pinup pose is inscribed, "To Henry with love, Maude. Taken May 1, 1946." Of course, his lack of intimate relationships may have resulted from his severe epilepsy and the antiepileptic drugs he was taking. Knowing that he could have a seizure at any moment must have made him extremely self-conscious in social situations; the potential embarrassment of having a convulsion during a date, or of slipping into a medication-induced nap, could be enough to discourage anyone from dating.

One of the greatest challenges to understanding who Henry was, and how he perceived his world, was that his memory of his life prior to the surgery was highly imperfect. He certainly suffered from *anterograde amnesia*—he could not remember events and facts that occurred after his brain was damaged. But he also experienced *retrograde amnesia*—he could not retrieve unique events he had encountered before the brain damage occurred.

Studying retrograde amnesia poses greater challenges than studying anterograde amnesia. To test for anterograde amnesia, all a researcher has to do is give a patient some items to remember—a picture, a sentence, a story, a complex drawing—and test later to see whether the patient has retained the information. On the other hand, studying retrograde amnesia is more difficult because it is challenging to figure out what information people had stored in the past. For this reason, researchers often personalize tests by using specific events and facts that are unique to the patient's own life and knowledge.

In 1986, two memory researchers at Boston University described a single case study that cast a spotlight on retrograde amnesia, an area of research that most previous studies of memory-impaired patients had neglected. The study addressed whether retrograde memory is equally affected across all time periods, or whether information stored decades before the onset of amnesia is more resilient than that stored closer to the disease onset. The researchers' carefully crafted experiments underscored the importance of having information about a patient's past knowledge. Using tests created expressly for their patient P.Z., they uncovered the relative sparing of *remote memories* coupled with the extensive loss of memories closer to the onset of his amnesia.[7]

P.Z., an eminent scientist and university professor, was diagnosed with alcoholic Korsakoff syndrome in 1981 at age sixty-five. He had both profound anterograde and retrograde amnesia. Because P.Z. had been a prolific writer, the memory researchers who studied him had an excellent grasp of what he had known before he became amnesic. He had written an autobiography before the onset of his brain damage, and when researchers tested his ability to recall events he had described in his book, he performed poorly overall. Intriguingly, his performance was uneven: he was more likely to give the correct answer about events that occurred further back in time. He remembered his early childhood well, but could answer almost nothing about the few years before the advent of his amnesia. As this and other studies have shown, distant long-term memory is less vulnerable than recent long-term memory.

To study this phenomenon in greater detail, the researchers worked up a list of famous scientists—seventy-five researchers whom P.Z. had known personally and in many cases cited in his own work. These scientists had achieved prominence during different time periods. In one experiment, the researchers showed P.Z. the scientists' names, one at a time, and asked him to identify each scholar's main field of study and specific scientific contributions. P.Z.'s worst scores were for the colleagues whose careers peaked after 1965, demonstrating that his retrograde amnesia swallowed the fifteen years prior to the onset of his amnesia in 1981. His best scores were for the period before 1965. We do not yet understand why patients like P.Z. experience this pattern of memory loss.

Henry also experienced retrograde amnesia, although it took decades for us to realize the true nature of what he had forgotten. The results of our experiments put him at the center of a scientific debate on the role of the hippocampus in *autobiographical memory*. From studying Henry's amnesia in greater detail, particularly his retrograde amnesia, we learned much about how the mind stores and retrieves different kinds of memories. The brain uses separate processes for retrieving *personal episodic knowledge*, such the morning your teacher promoted you to the top reading group, than it uses for *personal semantic knowledge*, such as the name of your elementary school. Our studies of Henry over half a century were crucial to this discovery.

At first, Scoville and Milner believed that Henry's amnesia was fairly straightforward—he could not remember any new information after the surgery, and had lost significant memory from the period immediately prior to his surgery, but had clear recall of things that had happened earlier in his life. In 1957, they reported that Henry had "a partial retrograde amnesia, inasmuch as he did not remember the death of a favorite uncle three years previously, nor anything of the period in hospital, yet could recall some trivial events that had occurred just before his admission to the hospital. His early memories were apparently vivid and intact." Similarly, in June 1965, a neurologist noted that Henry had a partial deficit in his knowledge of events that had occurred during the year prior to surgery. For example, he consistently confused a vacation he took a month and a half before the operation with one taken two months before the operation. The neurologist further documented memories that Henry retained—events that took place two or more years prior to the surgery, relatives and friends whom he knew preoperatively, and skills and abilities he once had. In 1968, based on information from Scoville's office and from unstructured interviews with Henry and his mother, we reported no change in Henry's capacity to recall remote events antedating his operation, such as incidents from his early school years, a high-school girlfriend, or jobs he held in his late teens and early twenties. His memory seemed vague for the two years immediately preceding his operation, performed when he was twenty-seven years old.[8]

As remote memory testing became more standardized and sophisticated, we realized that our early impressions were incorrect. From 1982 to 1989, my colleagues and I introduced objective tests to probe Henry's memory for different kinds of preoperative and postoperative information. The first kind was public knowledge—famous tunes (such as "Cruising down the River" and "Yellow Submarine"), information about widely known historical facts (*What agency controlled rationing and price support during World War II? In what Latin American country did President Johnson intervene by sending troops?*), and famous scenes (Marines raising the United States flag on Iwo Jima; Neil Armstrong on the moon). These tests intermixed items that Henry encountered before his operation with those that occurred after. We found that for public events that occurred during the 1940s through the 1970s—that is, both prior to and after his surgery—he was surprisingly accurate when choosing from four options. For instance, when asked, "When Franklin Roosevelt ran for a third term, his opponent was? . . . ," Henry correctly recognized Wendell Willkie. When asked, "Which world leaders did President Carter convene with at Camp David?" he correctly identified Begin and Sadat.[9]

Our test results revealed that Henry, despise his amnesia, could recognize some high-profile historical figures and events encountered after his operation. How do we explain this obvious indication of declarative memory in a man who for all practical purposes did not remember anything? We turned to his lifestyle for answers. He spent considerable time watching television and reading magazines, providing many opportunities to encode information about current events and celebrities. This repetition established representations—symbolic codes of information—in his cortex that were sufficient to allow him, when tested, to say whether he had encountered that person or event before. On the Famous Scenes Multiple Choice Recognition Test, Henry had to select one answer out of three choices—one correct and two incorrect. For example, when he was shown the image of Marines raising the United States flag on Iwo Jima, he chose from among three real-life events—Iwo Jima, South Pacific; Hanoi, Vietnam; and Seoul, Korea. He was also asked to pick one of three dates—1945 (thirty-nine years ago), 1951 (thirty-three years ago), and 1965 (nineteen years ago). When tested this way, with postoperative scenes, Henry often chose correctly. The memory traces, built up gradually through daily exposure to the media, sparked a sense of familiarity sufficient to support Henry's recognition memory. He was much less successful, however, when the examiner gave him a harder task—recalling the subject and date of the famous scenes (Iwo Jima, 1945) without any cues. A recall test like this is more challenging for all of us than recognition test because we have to dredge up the answers on our own. Henry's problem was disproportionately greater for postoperative years than the difficulty experienced by healthy participants. Although he scored normally for events from the 1940s, he was impaired for those from the 1950s through the 1980s.[10]

We also tested Henry's autobiographical memory. We asked him to relate personally experienced events cued by each of ten common nouns, such as *tree*, *bird*, and *star*. He could choose a memory from any time period in his life. We scored his responses on a scale ranging from zero to three, depending on how specific they were with respect to the time and place of the memory recalled. The participants earned a score of *three* for a memory that contained a specific autobiographical event that incorporated the stimulus cue (*bird*), was specific in time and place, and was rich in detail. For example, a participant might have said, "On my twenty-first birthday, I went to Las Vegas and stayed in a hotel that had green and red parrots in the lobby." They received a score of *two* for a memory that contained a specific autobiographical event that incorporated the stimulus cue but lacked specificity of time and place and showed poverty of detail: "I used to watch birds at a lake near my parents' house." They earned a score of *one* for a memory with autobiographical content but lacking the stimulus cue and specificity: "I used to enjoy bird watching." We gave participants a score of *zero* for no response or a general statement with no autobiographical reference: "Birds fly around."[11]

The results were telling. Henry drew his memories for personal events entirely from the period more than forty-one years prior to testing—when he was sixteen years old, eleven years before his operation. His surgery eradicated the most recent preoperative memories but spared the most distant. These results from the mid 1980s showed that Henry's retrograde amnesia was temporally limited, but the duration of the deficit was much more extensive than was reported in the 1950s and 1960s. People with anterograde amnesia, including individuals with dementia, often have lengthy retrograde amnesias, and they can remember events from their youth more clearly than events that happened just before the onset of their memory impairment. The shorthand for this phenomenon is "last in, first out."[12]

Subsequent advances in the design of remote-memory experiments gave us two new tools. The first, a more sensitive autobiographical memory interview, assessed our participants' ability to re-experience single events from specific times and places, including the recollection of contextual details. The second, a companion public-events interview, asked about the context of events that occurred in particular places at particular times. We carried out new studies in 2002 to assess Henry's memory for events in his distant as well as recent past. We began these experiments with a clue from my 1992 interview that he lacked any episodic, autobiographical memory.[13]

In this interview, I asked him, "What is your favorite memory of your mother?"

"Well, I—that she's my mother," he said.

"But can you remember any particular event that was special—like a holiday, Christmas, birthday, Easter?"

"There I have an argument with myself about Christmastime," he said.

"What about Christmas?"

"Well, 'cause my daddy was from the South, and they didn't celebrate down there like they do up here—in the North. Like they don't have trees or anything like that. And, uh, but he came north even though he was born in Louisiana. And I know the name of the town he was born in."

Henry's narrative began with Christmas, but as he continued he distracted himself, forgot the question, and ended on a different topic. Over years of interviews, Henry could not supply a single memory of an event that took place with his mother or his father. His answers were always vague and unvarying. When asked about a major holiday, most of us can recount vivid moments, replete with the sensory details that make memories indelible. Henry focused instead on sorting through facts, using his general knowledge about his family and his upbringing to try to construct a response.

This study, a breakthrough in our evaluation of Henry's memory, showed that his recollections from the time before his operation were sketchier than initially believed. He could conjure memories that relied on general knowledge—for example, that his father was from the south—but could not recall anything that relied on personal experience, such as a specific Christmas gift his father had given him. He retained only the gist of personally experienced events, plain facts, but no recollection of specific episodes.[14]

In October 1982, we had an excellent opportunity to explore, in a natural environment, Henry's memories from his preoperative life. I learned that his high school class was having its thirty-fifth reunion at the Marco Polo Restaurant in East Hartford. Neal Cohen, a postdoctoral fellow in my lab, and I obtained permission from the Bickford staff to take Henry out for a night on the town. We drove to Windsor Locks to escort him to the party, and on arrival found Henry all dressed up and eager to go.

The restaurant was crowded, with roughly a hundred people, classmates and their spouses, attending the reunion. Several of Henry's classmates remembered him and greeted him warmly; one woman even gave him a kiss, which he seemed to relish. As far as we could determine, however, Henry did not recognize anyone by sight or name. He was not alone in this predicament, however. One classmate confided to us that she did not recognize a single person at the event; unlike many of the attendees, she had moved away from Hartford and had not socialized with any of her classmates for many years.

Nor had Henry, of course—so we could not tell how much of his inability to recognize his classmates was a result of having no contact with any of them for thirty-five years, and how much was due to amnesia. Still, even if people's faces did not look familiar, he should have had glimmers of recognition when he saw their nametags. He might have said, "Danny McCarthy—I remember you from homeroom!" or "Helen Barker—I sat next to you in English, and you helped me with my homework." But he did not, suggesting that a significant chunk of his high-school memories had been erased.

As we studied Henry over the years, we learned that his deficit was specific to his autobiographical memory—although he was unable to retrieve unique life experiences, he remained capable of recalling public events with considerable clarity. So, for instance, he could talk about the stock market crash in 1929 (when he was three years old), Teddy Roosevelt's leading the charge in the battle of San Juan Hill, FDR, and World War II. When it came to personal information, however, his deficit was extreme. For example, he could recount a general picture of drives along the scenic Mohawk Trail in Massachusetts with his parents, but could not provide details of a specific event that occurred on a particular trip. He recalled facts but not experiences.

A giant in the cognitive science world, Endel Tulving, provided the theoretical breakthrough that helped us understand the distinction between the kinds of information Henry could and could not retrieve. In 1972, Tulving proposed two major categories of long-term memory: *semantic memory*, our store of facts, beliefs, and concepts about the world, and *episodic memory*, the unique events in our personal lives. Semantic memory is not linked to a particular learning experience—for example, I do not know when and where I learned that Paris is the capital of France. Unlike semantic memory, episodic memory records the flow of events in time and allows us to reflect on our mental representations in terms of what the event was, when and where it occurred, and whether it preceded or followed other events. We can vividly remember the details surrounding the phone call we received telling us that we were hired for the job, and to-day we can re-experience this exclusive event because of our capacity to mentally travel back in time. Henry was unable to make that voyage.[15]

What about his personal semantic knowledge—his factual memory for the people and places that pervaded his early life? Henry's old family photos, which Mrs. Herrick had given me, captured happy times—a wedding, catching a big fish, and celebratory family dinners. Here was physical evidence of his personal story. In 1982, I selected thirty-six of these photos for a test of Henry's childhood memories, interspersing these shots with an equal number of old photographs of my own family. I was not in any of these photos. I made slides of the photographs, pro-jected them onto a screen in the lab one at a time, and asked Henry whether he recognized the people in each picture and when and where the picture was taken. In one photo, Henry and his father were posing in front of a statue of a Native American on the Mohawk Trail; Henry, age twelve, wears shorts, glasses, a white dress shirt, and a tie, and is looking at the camera with his hands behind his back. His father, tall and lean, also wears a dress shirt and a tie, but is in long pants. He strikes a jaunty stance, looking off in the distance with his hands on his hips and one leg in front of him.

"Do you recognize these people?" I asked Henry.

"Well, yeah, one's me."

When I asked him which one, he replied, "The smallest one. And the other seems like my father, and it was taken—I think of the Mo-hawk Trail right off. There I have a question with myself, though—is that a statue? Well, I know that's a statue in the background—of an Indian. But I wasn't sure of that mountain that's in the background, further back."

When I asked him when the picture was taken, he said "about '38, '39 . . . '38. I said it right the first time." Although I do not know the exact date of the picture, his conservator and I guessed that he was about twelve years old, dating the photo to 1938, so his answer was likely correct.

Henry recognized the people in thirty-three of his thirty-six family photos. Equally significant, he did not recognize anyone in my family photographs. Only three of Henry's family photos evoked no memories for him. One was of distant relatives at a dinner table, and Henry was not in that picture. He said that the little boy looked familiar but did not know his name, where the photo was taken, or when. It is possible that he had not spent a lot of time with that family. In a photo of Mrs. Molaison and Henry celebrating his fiftieth birthday, he recognized his mother but not himself, seen in profile view. Here, Henry's error is difficult to explain because he did recognize himself in the other photos. The third photo he missed, taken after a hurricane, showed the exterior of his aunt's house with the roof missing. The house was in Florida, so Henry may never have been there or seen the photo. In every other case, he had specific knowledge of who and what was portrayed in his family photos. I was impressed that he recognized a landmark in one of my family photos (a shot of my mother holding my daughter in Gartford's Elizabeth Park, with seven ducks at their feet and a lake and trees in the background). He correctly identified the park, which he must have visited growing up in Hartford, and he was able to tease apart what he knew from what he did not know in this photo.

Henry's performance reflected his personal semantic knowledge—where he came from, his family history, and his own past. He had a general sense of identity. But, his deficient autobiographical knowledge—personal, unique episodes—meant that his self-awareness was significantly limited.

Memory researchers have long wondered why retrograde amnesia often seems to unfurl backward in time from the onset of anterograde amnesia, so that the distant past is most vivid and recent years fade away. One theory, the Standard Model of Memory Consolidation, posits that memories are consolidated—become fixed—over a long period that could last from months to decades. Psychologists Georg Elias Müller and Alfons Pilzecker at the University of Göttingen first proposed this theory in 1900. In the mid 1990s, neuroscientist Larry Squire and his colleagues at the University of California, San Diego, adopted the Standard Model as the centerpiece of their thinking about retrograde amnesia. According to this theory, the brain needs the hippocampal system during the early stages of consolidation to store and recover memories, but over time this system becomes unnecessary as areas in the frontal, temporal, parietal, and occipital lobes take over the responsibility for maintaining all memories over the long term. Once the memories are solidly stored, they no longer rely on the hippocampal system to be accessed again. In short, the hippocampal network plays only a temporary role for all forms of memory. According to this model, more recent memories are lost in cases of amnesia and dementia because these newer memories have not been fully consolidated and still depend on the hippocampal system.[16]

The flaw in the Standard Model is that it assumes that all memories are processed in the same way, whether they consist of general knowledge of the world (semantic) or personal experiences (episodic). Henry's recollections pointed to an important distinction between the two types of memory: he could remember facts he had learned before his operation, but when asked to recount what happened in his personal life at specific moments in time, he struggled. Declarative memories are not all processed in the same way. Through studying Henry, we learned that the ability to store and retrieve autobiographical memories that are detailed and vivid always depends on the hippocampal system, whereas remembering facts and general information does not.[17]

A better model for understanding Henry's retrograde amnesia emerged in the late 1990s, when neuroscientists Lynn Nadel and Morris Moscovitch introduced the Multiple Trace Theory of Memory Consolidation. Although Henry's case did not contribute to their original thinking, our findings with Henry provided strong support for their views. This theory draws on Tulving's proposal that we process facts differently from unique experiences. Like the Standard Model of Consolidation, the Multiple Trace Theory recognizes that semantic memories—world knowledge—can eventually become independent of the hippocampal system because they do not require us to remember the context in which we learned them or to connect and relate pieces of stored information. For instance, we do not remember that we learned the year of Columbus's first voyage to the Americas while sitting in the back row of a second-grade classroom; we simply remember *1492*. Remembering how we celebrated our twenty-first birthday, however, requires accessing the specifics of when, where, and what happened. According to the Multiple Trace Theory, it is possible to retrieve facts—like *1492*—without the hippocampal system, but it is not possible to retrieve any unique experiences—like the birthday celebration—unless the hippocampal circuits can communicate with the cortical circuits. Evidence from Henry's case favors the Multiple Trace Theory over the Standard Model of Consolidation because it indicates that traces of autobiographical events forever rely on the hippocampal system for retention and retrieval.[18]

The Multiple Trace Theory offers an alternate explanation for why people with amnesia are more likely to remember early experiences than later ones. According to this theory, neural processes in the hippocampus provide pointers, or an index, to all the distant sites in the cortex where memories of our experiences are stored. Think of this process in terms of a visit to a local library, where we look up a subject such as "birds of the Caribbean" in the card catalog, and then peruse the shelves to find the books. Each time we activate a memory trace by reminiscing about it or attaching a new piece of information to it, we create a new pointer, a new entry in the card catalog, to that memory. So, our memory of an exciting phone call with a job offer may be linked to multiple pointers within our hippocampal system through a web of associations from the times we have relived the moment or told other people about it. In this model, earlier memories have had a chance to knit themselves more firmly into the brain by accumulating these pointers over time. Retrograde amnesia hits newer memories harder because they are linked to fewer of these anchors and are, therefore, more vulnerable to being wiped away.

The controversy between the Standard Model of Consolidation and the Multiple Trace Theory focuses on autobiographical memory. Unlike semantic or general memory, autobiographical memory is episodic and rich in detail, and specifically includes what researchers call *experience-near details*—the specific sounds, sights, tastes, smells, thoughts, and emotions that accompany a unique event. The Standard Model assumes that autobiographical memories depend on a functioning hippocampus only for a limited time; after that, they become independent of the hippocampus and are stored in the cortex. This theory would predict, therefore, that Henry's autobiographical memory for his preoperative years was intact.

In contrast, the Multiple Trace Theory posits that retrieval of autobiographical episodes always requires hippocampal engagement. According to this view, the hippocampus is a memory index or pointer to the cortical areas that store the sensory and emotional qualities that make up episodic events. If the Multiple Trace Theory is correct, then Henry's ability to recollect autobiographical events in his early life would be impaired.

Henry's reports of preoperative autobiographical episodes were sparse; I am aware of only two. He related the first to Brenda Milner in the 1950s when he described a significant event that occurred when he was ten. "I can remember the first cigarette I ever smoked. It was a Chesterfield; I took it from my father's cigarettes. I took one mouthful, and did I cough! You should have heard me." For decades, this was the only autobiographical memory Henry conveyed to us.[19]

It was not until 2002 that we learned about a second autobiographical experience from Henry's past. At that point, we were systematically examining his autobiographical memory using a newly developed structured interview designed to elicit personal details. Sarah Steinvorth, a postdoctoral fellow in my lab, devoted several sessions to Henry's interviews, during which she asked him to describe an event from each of five life periods: childhood, teenage years, early adulthood, middle age, and the year prior to testing. She then asked him for as many details as possible about that event. If he had trouble thinking of an event, she would help by suggesting typical life events, such as a wedding or moving to a new house. The key to this study was patience and persistence. Steinvorth might have spent half an hour suggesting possible events to elicit a memory before moving on to another time period. She also discouraged Henry from choosing an event he had recounted repeatedly in the past. (We all have stories like these—events that we have retold so many times that we eventually narrate them blandly, without vividly reliving the wealth of sensory experiences.) As she questioned Henry, Steinvorth talked him through the different time periods of his life, and he struggled to produce any details. When she asked him for an event in his childhood, for instance, he mentioned falling in love with a girl whose father was a police captain, but was unable to describe a specific episode—something that occurred at a particular time and place—associated with that experience.[20]

Then one day, Henry delighted Steinvorth with an astonishing narrative.

"Can you think about one specific event, something that lasted several hours, from your early childhood up to age eleven?" she persisted. "Can you come up with something like that?"

"No, I can't," Henry said.

"Would you like to go to another time period, and see if you can come up with anything from that time period?"

"It would be better, in a way," he agreed.

"Okay. Let's try that. So maybe you can think of something—can you think about one specific event you were personally involved in that happened when you were between eleven and eighteen years old?"

He was silent for a while, so she repeated the question, but he still did not respond.

"Henry, are you tired? Would you like to have a break? Or are you just—"

"I'm trying to think."

"Okay, I'm sorry. I didn't want to interrupt you."

"I think of the first plane flight."

"Say that again."

"The first plane flight."

"Your first plane flight?"

"Yeah."

"Tell me about it."

Henry proceeded to describe, in great detail, the experience of a half-hour "sky ride" in a single-engine airplane when he was thirteen. He accurately described the layout of the Ryan aircraft, its instruments, steering column, and spinning propeller. He and the pilot sat side by side, and at one point, the pilot allowed him to take the controls; he recalled having to stretch his legs to reach the foot pedals. As Steinvorth questioned him, he remembered that it was a cloudy day in June, and as they flew around Hartford, he could see the buildings along Main Street. As they approached the airport to land, they passed over a cove where boats were docked. For the second time since his operation, a memory from the past was replete with experience—near details, the excitement of a specific event, complete with landmarks, colors, and sounds.

I was amazed when I read the transcript of Steinvorth's interview; I had no recollection of having heard this story before, but as it turns out, I was mistaken. In the course of writing this book, I reread an interview we conducted with Henry in 1977 while we were putting electrodes on his scalp for an all-night sleep study. He was chatting casually with the researcher when he related a similar narrative:

> And Brainard Field. Well, I know that too. Both before they had the transports in there, and when there were just private planes there. Why I remember is because in '39, that's where I went up. When I flew. Yeah. In an airplane. And I'm always glad I handled it too. That was something right there. 'Cause I know my mother and my dad were both afraid of airplanes. And when I went up, it was just before I graduated. It was because I was *going* to graduate that I was able to go up—about two and a half bucks. 'Cause the guy who took me up then was from—private pilot from Rockville—and he worked for the [unintelligible]. I got a little extra ride.

When I rediscovered this additional piece of evidence, obtained twenty-five years earlier, I was convinced that Henry's memory of the plane ride was genuinely autobiographical, and not a confabulation.

During the 2002 interview, Steinvorth asked Henry if he could remember any other specific events from his past. Although he came up with some promising attempts, none approached the vividness of the plane ride. He talked about taking a train trip with his mother when he was about seven, but simply recited facts, not unique episodes. They boarded the train in Hartford, switched trains in New York, and rode the second train all the way to Florida. He recalled sleeping on the top bunk while his mother took the lower one, and eating meals on the train. He also remembered visiting Canada with his parents while still in grammar school. He talked about milking a cow during that visit, and when Steinvorth quizzed him about the details, he could offer some information—sitting on a stool, being inside a barn with about twenty other cows, having to pull one teat and then the other. But again, despite her nudging, he could not paint the picture in a way that distinguished it as a genuine reliving of the experience rather than as a general description of milking a cow. It did not qualify as an autobiographical memory.

Henry's intense memories of his first cigarette and his airplane ride stood in marked contrast to his indistinct memories for the rest of his preoperative life. His ability to vividly recall these two extraordinary episodes is grounded in the powerful emotions he felt during both autobiographical experiences. Steinvorth asked Henry to rate his plane ride from one to six in terms of how much his emotional state changed during the flight, how personally important it was to him at the time of the flight, and how important it was to him at the time of testing. Henry rated his emotional state during the flight as six—a "tremendous emotional change." He rated the personal importance of the experience at the time of the flight, in retrospect, as five and at the time of testing as six. The memory traces of this unique experience stuck in Henry's brain because its salience and emotional significance strongly activated brain areas that support vivid encoding and storage of emotional information—his hippocampus, prefrontal cortex, and amygdala. This activation would have occurred not only at the time of the actual ride but also every time he told his friends about it afterward. Over time, the thrilling memory became more and more robust, a rich representation available for retrieval decades later.[21]

In contrast to his difficulty recounting specific episodes from his past, Henry consistently performed much better on tests of his remote semantic memory, general facts about the world. In one such test, Steinvorth asked him to focus on the public event itself rather than on his personal experience when he learned about the event—in other words, the semantic, not the episodic, knowledge. She gave him a list of cues and asked him to recall a specific public event, such as a major crime or the wedding of a celebrity, during different time periods in his life. Henry could recall public events from each of the preamnesic time periods. For instance, he chose to talk about "a big accident," and proceeded to describe the 1937 *Hindenburg* disaster in some detail. His ability to summon this type of general information lent further evidence to the idea that autobiographical and semantic memory are stored and retrieved in different ways. His intact memories of public events provided compelling support for the conclusion that his autobiographical memory deficit could *not* be attributed to a general failure in retrieving, recalling, or describing a detailed narrative structure.[22]

Scientists continue to debate the merits of the competing Standard Model of Consolidation and the Multiple Trace Theory. Our own results with Henry are consistent with the Multiple Trace Theory: Henry's ability to remember semantic information from his early, preoperative life remained strong, but his damaged hippocampus prevented him from remembering almost all autobiographical events. The two memories he could elicit—the first cigarette and the plane ride—were startling exceptions that revealed two outstanding moments in his life.

In an attempt to explain the nature of Henry's amnesia using the Standard Model, Squire raised two issues. First, he suggested that Henry's inability to retrieve preoperative autobiographical memories might have resulted from age-related disease, basing this suggestion on abnormalities in Henry's 2002–2004 brain scans. We can rule out this explanation, however, because we have evidence of this autobiographical memory vacuum in my 1992 interview with Henry, in which he not could retrieve an episodic memory related to his mother or his father. At that time, his brain did not show any age-related abnormalities.[23]

Squire's second explanation for Henry's lack of preoperative autobiographical memories was that Henry may have had autobiographical memories soon after his operation, but these memory traces withered away over time. If this reasoning is correct, then the same logic should apply to his memories of public events. They also should have dissipated over the years, but they did not. Henry's normal performance on the Public Events Interview showed that he could vividly recall public events from the very same years when his autobiographical memories were missing. His semantic-memory performance was spared, but his episodic, autobiographical memory was impaired. Based on our experiments, I stand by the view, in agreement with the Multiple Trace Theory, that we need a functioning hippocampus to re-experience unique moments in our past, regardless of how long ago they were acquired. An increasing number of studies also support this theory.[24]

A critical question in weighing the Standard Model of Consolidation against the Multiple Trace Theory is whether amnesia affects episodic, autobiographical memory and semantic memory differently. The two theories make clear predictions, and Henry's case exposed the difference between them. His test results, supporting the Multiple Trace Theory, taught us that the network of brain areas that supports the retrieval of remote autobiographical information is distinct from the network that sustains the recovery of remote semantic information. The former is impaired in amnesia, while the latter is not. Medial temporal-lobe structures are engaged in the initial encoding, storage, and retrieval of both kinds of memories. Then, during the process of consolidation, semantic memories become permanently established in the cortex, while episodic, autobiographical-memory traces continue to depend on medial temporal-lobe structures indefinitely. Thus, as far as we know, the removal of this tissue from Henry's brain left him with only two autobiographical memories.

In the late 1970s we did not know how important sleep is for memory consolidation, and we did not understand its central role in neural plasticity. At that time, we knew little about the neural basis of dreams, and studies of the cognitive neuroscience of dreaming did not exist. What we did know was that there is a relation between eye movements and different stages of sleep, and between different stages of sleep and dreaming. Armed with this basic knowledge, we set out to examine the effect of Henry's massive medial temporal-lobe damage on his dreams. We were tantalized by the Freudian possibility that his dream reports would expose stored secrets that we had been unable to elicit through his conscious recollections. Would we get a glimpse of his unconscious wishes?

230

Knowing that Henry could recount the gist of experiences from his preoperative life, we wondered whether these memories would provide the main content for his postoperative dreams. Dreams are the product of our imagination, akin to mental imagery when we are awake. They are typically disjointed, weird, and fleeting—not narratives that make sense. Experiments with rats have shown that dreams have meaningful connections with our waking lives. We became curious about Henry's dream content, given that he could not remember what he had done the day before.[25]

In 1970, I asked the CRC nurses to quiz Henry about his dreams when they woke him up in the morning. His answers tended to be similar from one day to the next, even though the nurse who woke him up changed from day to day. The same nurse might have elicited the same dream report on successive days. On May 20, he said he was running or being carried over hills; on May 22, he said he was driving a truck over hills with farmers, in pursuit of cattle rustlers; on May 23, he was in hills, but there were no trees; on May 26, he said he was in the country near the ocean—hilly—"like Louisiana with a sharp drop-off"; on May 27, he was racing over a hilly field with young men, age about twenty, to reach a spot so they could rest and sleep; and on June 6 he was walking through green hilly country—no trees.

To get a clearer understanding of Henry's dreams, we designed an experiment to document what he dreamed about. Our goal in this 1977 study was to establish what dreaming was like without a functioning hippocampus and amygdala. We monitored his sleep patterns during the night using electroencephalography (EEG), a tool that records electrical activity produced by neurons firing in the brain. These recordings told us what stage of sleep Henry was in. Two students helped capture Henry's dreams by waking him up during REM sleep and non-REM sleep, asking him whether he was dreaming, and if so what his dream was about. Like healthy participants, Henry reported dreaming during both sleep states.

Were Henry's dream reports genuine or merely anecdotes that he created on the spot in order to oblige his interlocutors? I suspect the latter. Of course, if Henry's dreams—like those of most people—were based on his own experience, they would have to draw on preoperative events, because he had no pool of recent memories to fuel them. Henry's dream reports were highly realistic and lacked the disjointed, unreal quality of most dreams. His typical responses too closely resembled his waking accounts of events in his youth—watching cowboy movies, enjoying nature, and taking car trips along Jacob's Ladder Trail and the Mohawk Trail in western Massachusetts. He had a small repertoire of factual stories related to preoperative experiences. Instead of describing real dreams, Henry was likely doing his best to be a cooperative participant, and what he came up with was the gist of memories from his dimly lit past.

Here is an example from a night when a student woke Henry at 4:45 a.m., during a period of REM sleep:

Student: Henry?—Henry?

Henry: Yeah?

Student: Were you dreaming?

Henry: I don't know. Why?

Student: You don't remember anything?

Henry: Well, in a way I do.

Student: What do you remember?

Henry: Well, I was trying to figure out—a house in the country, and I can't make out just how it's laid out. And—believe it or not—I was dreaming that I was a surgeon.

Student: You were?

Henry: Yeah—a brain surgeon. And—'cause that's what I wanted to be—but I said no because I wore glasses. And I said, well, a little speck [of dirt] or something, and that person could go [die] if you were performing an operation.

Student: Uh huh.

Henry: This is about the thought that I had about being a doctor, a surgeon, a brain surgeon. And that was—the whole line—I mean when I thought of that particular line of surgery.

Student: And you said something about being in the country? How's that?

Henry: In the country—performing operations or just being in the country too. Level area—I think of that—level area and wanted to know about that myself. Doubly, in a way, because I know that Dad had been brought up down south and it was flat down there. And I of course was brought up in Connecticut, and I milked a cow up in Canada. And—

Student: Did this all happen now?

Henry: No, this was reality.

Student: Oh, but then you weren't dreaming?

Henry: I was dreaming about the—putting it all together.

Henry's immediate report of "a house in the country" was possibly an actual dream, but here we run up against the problem of his limited span of memory. For his dream reports to be convincing, they would have had to occur within the span of his immediate memory, roughly thirty seconds. After that, the dream content would have evaporated, and the rambling conversation that followed would have tapped into his older stored knowledge.

I have no evidence to support the conclusion that Henry did not dream, but if he did, his dream experience must have been unlike that of healthy people. Some of the brain areas normally engaged during dreaming were replaced in Henry's brain by fluid-filled spaces. For instance, the amygdala in healthy participants is very active during REM sleep, and the absence of this activation in Henry's brain likely altered his sleep patterns and ability to dream. Also, he sometimes had nocturnal seizures whose aftermath left him out of sorts the next day, but despite our best efforts, we simply do not know the particulars of Henry's nighttime events. In 1977, when we woke him to ask whether he was dreaming, he sometimes said, "Yes" and other times "No." This pattern of responses indicates that he understood and evaluated the question, and that he did not come up with any old narrative on the spot just to satisfy the researcher. Still, the nature and quality of his dream content remain a puzzle.[26]

Henry's impairment in autobiographical remembering, both preoperatively and postoperatively, limited his self-awareness. He delighted in telling us tales about his relatives and childhood experiences, but they lacked precise details. His unfolding autobiography was deficient in the rich array of sensory and emotional narratives that form the intricate tapestry of who we are. Without the ability to travel consciously back in time from one episode to another, he was trapped in the here and now. Given these limitations, it is appropriate to ask whether Henry had a sense of who he was. Was his self-awareness blunted by his amnesia?

When people hear about Henry's case, they frequently ask me, "What happened when Henry looked in the mirror?" If he could not remember anything since his late twenties, how did he adjust to seeing himself as a middle-aged and eventually an older gentleman? When Henry looked in the mirror, he never expressed shock or a lack of recognition; he was comfortable with the person he saw looking back at him. Once, a nurse asked him, "What do you think about how you look?" In his characteristic understated humor, Henry answered, "I'm not a boy."

In the lab, we once showed Henry pictures of complex scenes, and he recognized them weeks later, based on a sense of familiarity and without explicitly recalling having seen them. Perhaps his own image did not surprise him for the same reason. Henry saw his face day in and day out for years. We know that the brain contains a region in the fusiform gyrus—a section of the temporal lobe that was preserved in Henry—specialized for processing faces. We also know that areas in the prefrontal cortex become active when people view their own face. These intact networks in Henry's brain may have allowed him to perceive his own face as familiar, even as it changed, and to continuously update his mental self-image.

At the same time, Henry's factual knowledge of his appearance and physical state was full of gaps. When we asked him his age or the current year, he often missed the mark by years or decades. He believed that he had dark brown hair even after he was partially gray, and described himself as "thin but heavy," despite packing on pounds as he aged. He somehow reconciled memories of himself prior to the operation with his current appearance.

During the decades following Henry's operation, his universe changed in countless ways, but he was never shocked by these transformations. He unconsciously became familiar with new information in his environment as a result of repeated exposures day after day, which gave rise to slow learning over time—different from the way that healthy people learn about their world. During each encounter with his own face, with the people who took care of him, and with his environment, his brain automatically registered their features and integrated them into stored internal representations of objects and people. Otherwise, his gray hair would have surprised him, and he would have constantly wondered why he lived where he did, why the image on the television screen was in color, and what computers were. He somehow accepted these additions and innovations as they appeared in his life.

Being unable to establish new memories, Henry could not construct an autobiography as his life unfolded, and the narrative of his past was also sketchy. For many of us, our personal history is the most critical part of who we are, and we spend considerable time thinking about our past experiences and imagining how our stories will play out in the future. Our sense of self includes the story of our past and where we think we are going—our "to-do list." We may imagine that we will advance in our careers, start families, or retire to better climates. In the short term, we have a plan for what to accomplish today, which friends we will see this week, and what we will do on our next vacation. Henry's operation, in addition to depriving him of his declarative memory, prevented him from mentally traveling forward in time, in the short term or in the long term. He lacked the pieces to construct agendas for the next day, month, or year, and could not imagine future experiences. In 1992, when I asked him, "What do you think you'll do tomorrow?" he answered, "Whatever is beneficial."

Cognitive neuroscientists have called attention to the link between simulating future events and episodic retrieval. They identified a common brain circuit that is engaged in remembering the past and picturing the future. The process of imagining future events depends on medial temporal-lobe structures, the prefrontal cortex, and the posterior parietal cortex—the same areas critical for declarative memory. When we fantasize about our next vacation, we tap into long-term memory for details of past vacations and other knowledge. Remembering these past events and recombining them to create future scenarios requires the retrieval of information from long-term memory, and it is no surprise that amnesia interferes with this process. Constructing the future, like resurrecting the past, requires establishing functional connections between the hippocampus and areas in the frontal, cingulate, and parietal cortices. Without this network, Henry had no database to consult when asked what he would do the next day, week, or in the years to come. He could not imagine the future any more than he could remember the past.[27]

Eleven

Knowing Facts

The sharp contrast between Henry's lack of episodic, autobiographical memory for the years *before* his operation and what seemed to be normal semantic memory for those same years raised the question of whether his episodic and semantic knowledge *after* his operation were equally affected. We had a wealth of evidence that his postoperative episodic memory was profoundly impaired—the hallmark of amnesia. But was Henry's ability to acquire new semantic memories normal or deficient? To what extent could he learn and retain semantic information he encountered for the first time after his operation? We also wanted to explore how well the old semantic knowledge he had acquired before his operation fared during the years that followed. These questions drove numerous research projects in my lab.

Postoperatively, Henry had a normal attention span and could still talk, read, write, spell, and carry on a conversation, drawing on knowledge he had acquired preoperatively. He could retrieve the semantic information he absorbed before his 1953 operation because it was stored throughout his cortex, and he did not need his hippocampus to access it.

We were particularly interested in Henry's language capacities because we wanted to know whether he needed his medial temporal-lobe structures to maintain his old semantic memories—knowledge about the world that he had squirreled away before 1953. A key component of semantic memory is lexical memory—stored information about words, including their meanings and forms (singular versus plural). The overarching issue was whether established lexical memories are preserved in amnesia. We designed our experiments to address three questions: Do medial temporal-lobe lesions compromise the capacity to use already-learned (preoperative) lexical information? Do these lesions affect grammatical processing? As time passes, does long-term lexical memory decay?[1]

When Henry spoke conversationally, his speech content was measured and deliberate. My colleagues and I wondered whether the characteristics of his speech processing were the same as in other people. In 1970, a graduate student in our lab proposed a test that would challenge Henry's speech-processing mechanisms with linguistic ambiguities, sentences that have more than one meaning. Generally, there are three types of linguistic ambiguity—lexical, surface structure, and deep structure.

Lexical: *When a strike was called it surprised everyone.*
Surface structure: *A moving van out of control is dangerous.*
Deep structure: *Visiting relatives can be a bore.*

The student created sixty-five ambiguous sentences, including multiple sentences for each type of ambiguity, as well as twenty-five unambiguous sentences with one meaning (for instance, *Jim bought a parka at the ski shop.*)[2]

In the experiment, the student read the sentences aloud, and a prompt card in front of Henry asked, *Does the sentence have one meaning or two?* Normal participants could mentally restructure and correctly interpret the ambiguous sentences. Although Henry did not detect the presence of two meanings as often as our control subjects, when he did, he grasped all three kinds of linguistic ambiguity, including deep-structure ambiguity—as in *Racing cars can be dangerous.*

This study illustrated Henry's normal capacity to retain, over several seconds, the various elements of a sentence and their relation to another. He detected ambiguities less frequently than control participants because his short-term processing capacities became overloaded and his long-term memory failed to activate. Short-term memory can store a small amount of information temporarily, and although Henry's brain could do that, its limited capacity was insufficient to disambiguate some of the sentences.

Around the same time, a psychologist at the University of California, Los Angeles, reported a similar set of experiments. Working independently when he was a graduate student in our department at MIT, he created thirty-two ambiguous sentences and read them aloud to Henry, giving him two instructions: "Find the two meanings of the sentence as fast as possible," and "Say 'yes' and give the two meanings in the order you saw them." This researcher imposed a ninety-second time limit, so if Henry did not find two meanings during that window, the trial was counted as an error. Henry detected the two meanings in sentences with lexical and surface ambiguities on more than eighty percent of the trials, but when given sentences with deep-structure ambiguities, he scored zero. Contradicting our previous results, the California researcher contended that structures removed from Henry's brain—"the hippocampal system"—played a central role in language comprehension.[3]

This conclusion seemed incorrect to my team. We knew Henry to be conversational and improvisatory in talking with everyone he met. We also understood from decades of research dating back to the mid 1800s that language expression and comprehension were not localized to the hippocampal system. Many researchers, including myself, argued that language is mediated by multiple cortical circuits, primarily in the left hemisphere in most individuals, and not by the hippocampus or parahippocampal gyrus.

I decided to test the UCLA researcher's findings by administering his sentences to Henry myself, with important changes in procedure. I asked him to read the sentences aloud, but also to reread any sentence from which he omitted a word; I called his attention to omissions by putting my finger under words he had overlooked. Then, I asked him to say "yes" when he saw two meanings in the sentence and to describe them in the order in which he noticed them. Because Henry generally did everything slowly, I gave him unlimited time to interpret the sentences, instead of cutting him off after ninety seconds as the UCLA researcher had done. My results showed that Henry could indeed detect deep-structure ambiguities, provided he read the entire sentence without omissions, and provided he went at his own pace. Indeed, the fact that Henry could converse effectively—as he did every day with lab members and the CRC staff—attested to his ability to grasp the underlying meaning in sentences.

We continued to study Henry's linguistic abilities until nearly the end of his life. We did not believe, as the UCLA researcher did, that his bilateral medial temporal-lobe lesions impaired his appreciation of linguistic ambiguity or any other speech-processing capacities. To make our case definitively, in 2001 a graduate student and a postdoctoral fellow in my lab gave Henry a host of tasks to assess his store of word knowledge and his ability to use grammatical rules. After examining his performance on nineteen tests, we concluded that his trouble detecting ambiguities did not result from a fundamental deficit in word knowledge or grammar.[4]

Henry could readily name objects depicted in color pictures and line drawings. For instance, the researchers showed him cards with a picture and a word. Half the time, the word matched the item in the picture, and the other half, the word and the picture were unrelated. Henry could distinguish between the matching and non-matching cards with only a slight deficit. Likewise, he performed well on tests of basic grammar. For instance, he could give the plural forms of nouns and the past-tense forms of verbs, and could turn an adjective into a noun ("The man is stupid. In fact, his _____ is noticeable.") Henry scored as well as control participants when we asked him to listen to a sentence and determine whether it was grammatically wrong.[5]

Henry did, however, have difficulty on tests of language fluency. In one such test, we gave him a category, such as "fruits," and asked him to name as many examples of the category as he could think of in one minute. In another, we asked him to name in one minute as many words as possible beginning with the letter *F* then *A* then *S*. These letters sampled a range of difficulty. Based on the number of words available to choose from, *F* words are the hardest, *S* words are the easiest, and *A* words fall in between. On both of these fluency tests, Henry's scores were inferior to those of the nineteen control participants. Still, the results for all other language functions showed preserved lexical-memory performance (word knowledge).[6]

Henry's scores on measures of fluency were impaired—of that, there was no doubt. The most straightforward explanation for this limited performance was his lower socioeconomic status. Prior to his operation, he was by no means a highly verbal person, and his poor naming ability after his operation probably reflected a general lack of verbal skills. His working-class upbringing may have limited the development of his language processes. He never went to college, and his skills and interests as a young man tended toward technology and science. Language was not his strong suit. The letters Henry received from friends stationed overseas during World War II, with their frequent misspellings and grammatical mistakes, underscored our impression that language skills were not a priority in his social group. The bulk of our research ultimately indicated that overall, Henry's language abilities were consistent with his socioeconomic status and likely the same as before his operation.

Lab members and I found that during our informal interactions he could appreciate puns and linguistic ambiguities, such as words with double meanings. Henry rarely initiated conversations, but when we engaged him, he was always a willing, communicative, and entertaining participant. Once, when I said to him, "You're the puzzle king of the world," he replied, "I'm puzzling!"

Henry's operation did not affect most of his linguistic abilities because the many brain areas that support the production and understanding of language are outside the medial temporal-lobe region. Beginning in the late 1980s, functional brain-imaging experiments added a different dimension of information to the understanding of language. Two new

241

scanning tools, positron emission tomography (PET) and functional MRI, allowed researchers to observe brain activity while a healthy person in the scanner was performing various kinds of tasks that used words. Different technologies underlie PET and funtional MRI. For PET studies, the participant is injected with a radioactive tracer that is taken up by neurons—particularly those that are most active—and is detected by a complex X-ray machine. The analysis tools allow researchers to link discrete areas of activation to the specific cognitive processes engaged by the person in the scanner. Functional MRI relies on a different technology to link brain and behavior, one that uncovers areas of task-related activation using measures of blood flow. Functional MRI has largely replaced PET in cognitive neuroscience studies because it does not expose participants to radiation, and it gives a more precise picture of brain activity.

A 2012 review of 586 functional-imaging experiments synthesized results concerning the localization of activity related to heard speech, spoken language, and reading. This overview showed thirty-one areas of language-related activation in the cortex on the left side of the brain, as well as in structures under the cortex—the caudate nucleus, globus pallidus, and thalamus—and in two places on the right side of the cerebellum. Each area supports one or more aspects of language function, such as processing speech sounds, speech comprehension, speech production, processing written words, and converting spelling to sound. These cortical and subcortical areas are highly interconnected by white-matter fiber tracts, allowing them to communicate with one another efficiently. The right hemisphere also participates in language functions. A network in the right frontal and temporal lobes processes information about speech rhythms, accents, and pitch.[7]

Henry's unique situation enabled our team to again make an impact on cognitive neuroscience—this time in the area of language. Our study of his lexical and grammatical processing was the first extensive analysis of these capacities in amnesia. His data revealed a striking distinction within lexical memory between retrieval of previously acquired information (which was preserved) and new learning (which was not). Henry's test results showed clearly that medial temporal-lobe structures are not critical for the retention and use of preoperatively learned lexical information and grammar. He could spell familiar words, identify objects by name, match pictures with the corresponding words, and say where famous landmarks were located. This capacity to retrieve lexical information and use that information efficiently was rooted in the intact cortical networks that support language. By contrast, we learned from Henry's case that medial temporal-lobe structures are required for learning *new* lexical information, as we shall see in his inability to learn words that were not in his preoperative vocabulary.

How did Henry's preoperative semantic knowledge survive over time? Was he able to retain these memories as well as people without brain damage? In healthy people, semantic memory is less vulnerable to the ravages of time than is episodic memory. In fact, experiments conducted in the 1960s indicated that older adults often scored higher than young adults on tests of their general knowledge of the world. Of course, as you age, you have more opportunities to build up storehouses of words, concepts, and historical facts—as well as to reconsolidate information you have already learned. For instance, each time you hear or read the word *espionage*, you automatically access its meaning in your semantic storehouse and process it. In this way, you enrich your memory trace for *espionage* over time. This continual rehashing of information may explain why some semantic memories are indelible.[8]

We wanted to understand whether Henry had kept his preoperative semantic store intact in the same way that healthy older adults had, and whether he had shown consistent performance from year to year in retrieving word information. One of the advantages of studying him for decades was that we could compare his performance on the same IQ tests given over and over again. No studies had examined the stability of memory for words over time in amnesia, so we broke new ground when in 2001 we reviewed forty-eight years of Henry's test results.

Researchers in my lab analyzed Henry's test scores from twenty testing sessions carried out between August 24, 1953 (the day before his operation when his brain was still intact) and 2000. This analysis evaluated his performance on four subtests from a standardized IQ test: *General Information* (Who wrote *Hamlet*? On what continent is Brazil? How many weeks are there in a year?); *Similarities* (In what way are an eye and an ear the same?); *General Comprehension* (Why is it better to build a house of brick than of wood?); and *Vocabulary* (What does *espionage* mean?). We found that Henry's performance on these four tests was consistent across forty-eight years. His memory for facts, concepts, and words remained constant from the day before his operation through 2000, indicating that medial temporal-lobe structures were not critical for retaining and using the word knowledge and concepts he had consolidated before his operation. Importantly, the results showed that Henry's brain could maintain information it had already learned, without explicit practice. Because of his amnesia, he did not have the benefit of episodic learning, but he could still preserve word knowledge by engaging brain circuits outside the hippocampus, in his frontal, temporal, and parietal lobes. He believed that doing crossword puzzles helped his memory, and perhaps he was correct.[9]

This kind of retrospective study was possible only because we had been collecting detailed information about Henry's semantic knowledge for decades. A related goal of our research was to further examine his semantic memory as thoroughly as we had scrutinized his episodic memory. We wanted to go beyond the routine assessment provided by standardized tests, and make sure we did not leave any hidden corners of his memory unexamined.

In 1970, memory researchers in England proposed the idea that the memory difficulty experienced by amnesic patients boiled down to an abnormality in retrieval—they could store new memories normally but were simply unable to consciously bring them back. The researchers further argued that ostensibly forgotten information could be lured out of amnesic patients by giving them prompts. If that were true, then giving Henry prompts related to postoperative material should have catapulted his performance into the normal range. A 1975 experiment in our lab tested this proposal. A graduate student working with Hans-Lukas Teuber designed a famous-faces test using news photographs of public figures who had been eminent at various points in time from the 1920s to the 1960s. He first asked Henry to identify these people without any hints. If unsuccessful, Henry received two kinds of prompts to help him—circumstantial and letter. For instance, the circumstantial prompt for Alfred Landon was, "He was the Republican presidential nominee in 1936; he ran against Roosevelt and lost; he was also Governor of Kansas." If Henry still could not identify the public figure, he received phonemic prompts that provided an increasing number of letters in the first and then the last names of the public figures, moving from the initials and ending with almost the whole name. For Alfred Landon, the phonemic cues were "A.L., Alf. L., Alfred L., Alfred Lan., Alfred Land." When we compared Henry's semantic memory for the time periods before and after the onset of his amnesia, we found that he retained his memory for public figures from the years before his operation, but his memory for public figures from the period after was markedly inferior to that of the control participants. For the post-1950s public figures, the prompts were of little help. Henry had not successfully encoded, consolidated, and stored this semantic information; clearly his amnesia could not be dismissed as an instance of faulty retrieval.[10]

During the following decades, lab members updated this test and gave it to Henry in nine test sessions from 1974 through 2000. We then harnessed this huge dataset to determine whether he performed consistently across years of testing. When all the data were combined, we found that he was as good as, or better than, healthy control participants when asked about people who were famous in the 1920s through the 1940s but was woefully inferior to control participants for the 1950s through the 1980s. For example, without being prompted, he correctly identified Charles Lindbergh and Warren G. Harding from the 1920s, Joe Lewis and J. Edgar Hoover from the 1930s, and John L. Lewis and Jackie Robinson from the 1940s. After the 1940s, he was at a loss. He could not name Stan Musial and Joseph McCarthy from the 1950s, John Glenn or Joe Namath from the 1960s, Jimmy Carter or Princess Anne from the 1970s, or Oliver North or George H.W. Bush from the 1980s.

Still, it was apparent that Henry was storing some new information. With prompting, he was able to recognize a few public figures whose fame began after 1953, but the number of prompts he required was fifty percent greater than the average for the control participants. So, even with generous cueing, Henry's memories were difficult to tease out. Clearly, more postoperative material is stored than can be retrieved without extensive prompting, but if a general failure of retrieval were the basis of Henry's amnesia, then this deficit should have compromised the retrieval of *preoperative* material as well. The fact that it did not, reinforces the view that amnesia is rooted in an inability to continuously consolidate, store, and retrieve life's experiences.

Even though Henry's case provided proof that the medial temporal lobes were necessary to form both kinds of declarative memory—episodic and semantic—the idea was not accepted without controversy. Some researchers doubted that new semantic learning depended on medial temporal-lobe structures the way episodic memory did. In 1975, two clinicians in Toronto proposed that amnesic patients have damage to brain structures that support the acquisition and retrieval of episodic memories, but *not* to structures that mediate acquisition and retrieval of semantic memories. In 1987, one of them argued that amnesic patients should be able to learn new facts whenever retrieval of the facts did not depend on having explicit recollection of the specific event during which the patient learned the fact. This scientist predicted that patients like Henry could acquire general knowledge outside of conscious awareness, through nondeclarative memory circuits.[11]

In 1988, my colleagues and I put this view to the test by trying to teach Henry new vocabulary words. We wanted to see if he could learn the definitions of eight words that were in the English dictionary but were not in common use: *quotidian, manumit, hegira, anchorite, minatory, egress, welkin,* and *tyro.* These were all words, we suspected, that Henry would not have encountered preoperatively. Henry viewed these words on a computer monitor, one at a time, with a single definition. He read each word and its definition aloud. He then saw all eight definitions with one of the eight words below and had to select the correct definition for that word. If he answered correctly, the definition was removed from the choice list, and a new word appeared at the bottom of the screen. If he answered incorrectly, he was asked to select another definition. This procedure continued, word by word, until Henry selected the correct definition for each of the eight words. Control participants were, on average, able to match all of the words with the definitions in fewer than six trials. Henry, however, failed to learn these new vocabulary words in twenty trials.[12]

We persisted and tried to teach Henry the same words using two additional methods—giving a common single-word synonym for each of the eight words and filling in the blank in sentences in which one of the eight words was missing. Henry apparently knew the meaning of *tyro* (presumably he had learned it prior to his operation) because he selected the correct definition and synonym every time, and filled in the blank correctly on ninety percent of the trials. But he never mastered the meaning of any other word, providing unequivocal evidence that he could not learn, in a controlled laboratory setting, the meaning of any new words. In contrast, the control participants continued to absorb the novel words in fewer than six trials.[13]

Yet one could argue that these experiments did not mimic the natural way that people learn the meaning of new words in their everyday lives. Perhaps our laboratory setup was too artificial and did not sufficiently tap into Henry's true ability to acquire semantic information. On a daily basis, we are exposed to new words in a variety of meaningful and relevant contexts in numerous situations. We often see or hear these words in pursuit of a goal, so we are motivated learners. Picking up on this idea, another team proposed in 1982 that the decisive test of whether an amnesic patient could acquire new vocabulary words would be to take the person to a country where he did not speak or understand the language. The researchers predicted that the patient would learn the new tongue slowly, as a child would, while afterward forgetting that he had even been there. This would be a more natural learning environment than a laboratory because it would combine hearing, speaking, reading, and writing the language. Rather than trying to learn words as Henry did in the lab, the amnesic person in a foreign country might learn phrases and sentences in meaningful contexts, such as the bakery, pharmacy, coffee shop, or park. According to the argument, he might, with repeated exposure, be able to build up a rich mental representation of the language, consisting of its speech sounds, vocabulary, concepts, and grammar.[14]

We sensed that this theory was wrong, but in order to prove it we needed to see if Henry had incidentally acquired some knowledge of words that were new to the English language since his operation in 1953—words that he may have encountered in daily life. Even if he could not remember the definition, it was possible, for example, that he knew a new word was a real word without knowing its meaning. This intuition is common in healthy people.

We set out to assess Henry's knowledge of new words that had been added to the Merriam-Webster Dictionary after 1954, words he probably encountered after the onset of his amnesia. The test stimuli were words like *charisma, psychedelic, granola, Jacuzzi,* and *palimony*. They were intermixed with old words (*butcher, gesture, shepherd*) and pronounceable nonwords (*phleague, thweige, phlawse*). We wanted to know whether Henry considered the post-1954 words and the pronounceable nonwords to be legitimate words. Each trial began with the following question on the computer screen, "Is the following a real word?" Henry read the word and responded "yes" or "no." He was right if he answered "yes" for the legitimate words and "no" for the nonwords. He correctly said "yes" to ninety-three percent of the pre-1950s words (which was normal compared to the controls' ninety-two percent) and fifty percent of the post-1950s words (which was impaired relative to the controls' seventy-seven percent). His ability to categorize nonwords as nonwords was borderline normal—eighty-eight percent versus the control group's score of ninety-four percent. This relatively simple experiment strengthened the distinction between Henry's preoperative semantic knowledge, which was intact, and his postoperative semantic knowledge, which was severely lacking.[15]

To explore another dimension of Henry's semantic knowledge, we designed a test to measure how much he knew about public figures. A minimal expression of knowledge for the names of famous people would be the ability to recognize a famous name as such. That is what we asked Henry to do in a name-categorization task—"Is or was the following a famous person?" Henry answered "yes" or "no." The celebrities included movie stars, athletes, American politicians, foreign leaders, and writers. On this task, the names of people who became famous either before or after Henry's operation were intermingled with similar names chosen from the Boston area telephone directory. Henry matched the control participants in rejecting the non-famous names as famous, and he was slightly better than the controls in identifying famous personalities in the 1930s and 1940s (eighty-eight percent for Henry and eighty-four percent for the healthy controls). For the people who became famous in the 1960s, 1970s, and 1980s, and who were almost certainly unknown to Henry before his operation, his score of fifty-three percent was well below the controls' eighty percent correct. This pattern of results—intact memory of public figures for the preoperative period and impaired memory for such people from the postoperative period—reinforced the conclusion that Henry could not successfully store and retrieve factual information that had entered the world after the onset of his amnesia.[16]

But in spite of this impoverished test performance, Henry's semantic storehouse did retain tiny traces of his experiences during the postoperative years. We found tantalizing evidence that his amnesia was not absolute. His *recognition* memory was not totally wiped out. When we asked him to select the best definition for words and phrases among four choices printed in a test booklet, he recognized fifty-six percent of the definitions for pre-1950s words and phrases and thirty-seven percent for the post-1950s words. The latter score, though impaired, was better than chance (twenty-five percent). At the same time, his ability to *recall* these words and phrases was clearly compromised. He successfully fished up sixty-one percent for pre-1950s words and phrases but only fourteen percent for post-1950s words and phrases.[17]

Likewise, Henry's performance on the word vs. nonword task ("Is the following a real word?" *thweige*) and the name-categorization task ("Is or was the following a famous person?" *Lyndon Baines Johnson*) also demonstrated some, albeit scant, knowledge of post-1950s words and names. When the examiner asked Henry to define words that had entered the dictionary after he became amnesic, he usually did not know the answer. But rather than just say, "I don't know," he would harness his intellect and make an educated guess. Many of his responses construed literal meanings from bits of words and phrases. He defined *angel dust* as "dust made by angels; we call it rain," *closet queen* as "moths," *cut-offs* as "amputations," and *fat farm* as "a dairy." The brain areas in Henry's cortex that housed his semantic store for old words and names could not engage in learning new words and names because the critical interactions between these key cortical areas and intact hippocampal circuits were missing. For the most part, he could not consolidate new semantic memories. Returning to the question of whether an amnesic person placed in a foreign country would learn the language but forget having been there, it appeared from Henry's performance that even in a natural learning environment in which the individual heard, spoke, read, and wrote the language in meaningful contexts, he would be unable to expand his word knowledge with new entries. This deficit was part and parcel of Henry's profoundly impaired declarative memory.[18]

Extending this line of thinking, we conducted several experiments in the mid 1990s following up on Henry's lack of semantic learning for new words. The question that intrigued us was whether this deficit in acquiring novel information through declarative memory would carry over to *nondeclarative* memory, which we knew was intact. Although he could not consciously recollect new words, would he show *priming* with these words? Would he be able to process them normally through preserved nondeclarative-memory circuits? Specifically, we wondered whether Henry's priming performance would be different with words that came into common use after the onset of his amnesia and thus were novel to him—post-1965 words like *granola, crockpot, hacker,* and *preppy*—compared with old words like *blizzard, harpoon, pharmacy,* and *thimble.*

A graduate student in my lab created four repetition-priming tasks. Two of them assessed Henry's *word-completion priming*—one using words that were in the dictionary before his 1953 operation and the other using words that entered the dictionary after 1953. The two other tasks measured Henry's *perceptual identification priming*—one with pre-1953 words and the other with post-1965 words. All four tasks engaged Henry's nondeclarative memory circuits. In this experiment, each priming task had a study condition and a test condition.[19]

In the *study condition* for word-completion priming, Henry read words aloud as they were presented one at a time on a computer screen. The *test condition* occurred one minute later when he saw three-letter stems one at a time on the computer screen (*GRA-* for *granola*, *THI-* for *thimble*). Half of the word stems corresponded to words in the study list, and the other half to unstudied words. The examiner asked Henry to complete each stem with the first word that jumped into his mind. If he completed *THI-* to *thimble*, rather than to more common words that begin with *THI-* (*think, thin, thief, thick*), then he was showing the effect of prior exposure to *thimble*, the essence of word-completion priming. The word-completion priming score was the number of stems completed to studied words, like *thimble*, minus the baseline score of words completed by chance to similar unstudied words. Henry showed priming whenever he completed significantly more word stems to studied than to unstudied words.[20]

Perceptual identification priming also included a study condition and a test condition. During the *study condition*, we flashed words on a screen, one by one, for less than half a second and asked Henry to simply read each word aloud. In the *test condition*, Henry again saw words presented very briefly on the screen and again read each one aloud. He had seen half of the test words in the study session, and the other half were new words he had not studied. His perceptual-identification priming score was the number of correctly identified studied words minus the baseline score of correctly identified unstudied words. Henry again showed priming. When the words flashed briefly on the screen during the test, he was able to read more of the studied than unstudied words. During the study condition, each time Henry read a word, the experience left behind a memory trace that beefed up the representation of that word. Then, during the test, when he saw studied and unstudied words for less than an instant, he was more likely to read the studied words because of their enhanced mental representation.[21]

Henry gave us clear-cut results on these four priming tests. For the pre-1950 words—*blizzard, harpoon, pharmacy, thimble*—he showed normal performance on both kinds of priming. On stem-completion priming, he completed more stems to studied than unstudied words, and on perceptual identification priming, he correctly read more briefly presented studied than unstudied words. We found a different breakdown of results with the post-1950 words—*granola, crockpot, hacker, preppy*. Henry's perceptual-identification priming was still normal, but his word-completion priming score was zero. Why? Henry lacked the necessary preexisting semantic representations of the novel words in his mental dictionary for word-completion priming to occur. He did not need this representation for perceptual-identification priming because the task engaged low-level visual processes independent of language. This finding of impaired word-completion priming with novel words, but robust perceptual identification priming with the same words, told us that different mechanisms sustain each kind of priming—one disrupted in Henry and the other still functioning.[22]

Henry's differing results for these two nondeclarative-memory tasks—conceptual priming and perceptual priming—were important because they highlighted the fact that the two procedures activated brain circuits at different levels of information processing. Word-completion priming operated at the level of word knowledge stored in the temporal and parietal lobes. The word *granola* was unfamiliar to Henry and not represented in his mental dictionary—his semantic store. Consequently, when he read it in the study list, he did not have a representation available to benefit from the extra processing needed to complete the stem *GRA-* to *granola* in the test. He had no difficulty, however, completing *GRA-* to *grandmother* because it was already part of his preoperatively stored world knowledge. This experiment confirmed our previous work showing that Henry had normal word-completion priming when we tested him with words that were familiar to him.[23]

In contrast, perceptual-identification priming operated at the more elementary level of visual perception. Henry simply read *granola* aloud in the study list, and the related processing in his visual cortex was all he needed in the test to say the word aloud when it flashed on the screen. Henry showed comparable perceptual-identification priming for pre-1953 and post-1953 words because his visual cortex in the back of his brain, where these computations were performed, processed meaningful (*blizzard*) and non-meaningful (*granola*) letter strings in the same way.[24]

Amnesic patients, including Henry, characteristically show learning on nondeclarative memory tasks, such as priming. Our finding that Henry was impaired on word-completion priming with novel words was an exception. This deficit resulted directly from Henry's inability to consolidate and store post-1953 words. In this priming task, he read words like *granola, crockpot, hacker*, and *preppy* on a computer screen, but he had no entry in his mental dictionary for them to activate. He could not benefit from reading the words and was, therefore, unlikely to respond with *granola* when he saw *GRA-*. His performance showed that the circuits in his brain that supported this kind of priming were distinct from those that sustained his normal perceptual-identification priming with the same words. This division of labor occurs in our brains as well.[25]

Although Henry demonstrated over and over that he had lost his ability to make new semantic memories, to learn new facts and retain them, he occasionally astonished us by remembering things we never expected him to know. One day, in casual conversation, research associate Edith Sullivan asked Henry what he thought of when he heard the name Edith. She was flabbergasted when he said, "Edith Bunker." Edith Bunker was a fictional character from *All in the Family*, a TV show that began in 1971. The next day, she brought up the subject again and asked him: "What's the name of the star of the program?"

"Archie Bunker," Henry said.

She asked him what Archie Bunker called his son-in-law, adding, "it's not a very pleasant name."

After a long pause, Henry said: "Meathead."

These surprising and seemingly random new memories appeared from time to time like driftwood washing up from an empty sea, and they felt like small miracles to those of us accustomed to seeing Henry fail to remember. In the early years, Henry's mother often believed he was improving—she told us, "He knows things that he has no business to know." In hindsight, it is clear that Henry's amnesia was permanent, and these fragmentary scraps of memories were the exception not the rule. Compared with any normal person, Henry's ability to remember the experiences of his life was always abysmal. In 1973, he was not able to identify common names in the news at the time, "Watergate," "John Dean," or "San Clemente," despite the fact that he had heard them repeatedly on television every night. He did not know who the president was, but when told his name began with an "N," he said, "Nixon."

In July 1973, I asked Henry whether he could tell me about Skylab. He replied, "I think, uh, of a docking place in space." He also correctly said that there were three people in Skylab at the time, but immediately added: "But then I had an argument with myself, then was it three or five?" When I asked him, "What's it like to move around up there?" he answered, "Well, they have weightlessness—I think of magnets to hold them on metal parts so they won't float off away, and to hold them there so they can move around themselves and stay in one area, and they won't move away unvoluntarily [sic]." Henry accepted new technologies, such as computerized tests, without blinking, but he could not keep up with changing times in other ways—he once incorrectly stated that a hippie was a dancer. The world changed, but for the most part Henry was left behind.

The little islands of remembering that Henry treated us to over the years spurred us to investigate his semantic memory more intensely. We decided to again examine his knowledge of celebrities because it allowed us to take advantage of his extensive exposure to magazines and television through which he constantly took in information about famous people and noteworthy events. A 2002 experiment when Henry was seventy-six further characterized the depth of his semantic learning by probing for details of individuals who became famous after his 1953 operation. Our previous studies had not explored the depth of the material he was able to acquire postoperatively. The fragments of knowledge he had demonstrated in our prior experiments were sufficiently sparse that either declarative or nondeclarative learning could have supported them. (Two other groups of memory researchers had each shown that a severely amnesic patient, exposed to many weeks of training, could gradually learn and retain new semantic facts. The patients demonstrated their scant factual knowledge through performance but could not consciously remember the learning episodes. The learning was nondeclarative.)[26]

In our study, two graduate students addressed an ongoing controversy. Some researchers predicted that regions beyond the hippocampus proper could support some semantic—consciously accessible—learning. But another lab argued that amnesic patients' episodic and semantic memory circuits are equally impaired, so that semantic learning would be impossible in someone like Henry who had no episodic memory. We designed experiments to reveal the route through which we acquire semantic knowledge. The new theoretical issue that motivated our experiments was whether all new information enters your brain in the form of episodes and later becomes general knowledge. For instance, the first time you ate peach ice cream and discovered you loved it was at the beach when you were twelve years old. Over the years, you forgot that episode but continued to rate peach ice cream as your favorite. This fact began as an episodic memory and later became a semantic memory. So the question was, do all memories have to start like this, as episodes, or can they bypass your episodic circuits and enter your brain as semantic knowledge? Because the hippocampus is necessary for episodic learning, we could also phrase the question in this way—can semantic learning occur without a functioning hippocampus? Because Henry's hippocampal damage was complete and because he had virtually no episodic memory, he was the perfect case for testing this hypothesis.[27]

In our initial experiment, Henry heard the first names of famous people and was asked to quickly complete the name with the last-name that immediately came to mind. Because this task did *not* ask for a famous name—any name would do—implicit (nondeclarative) memory could support performance, automatically without awareness. For example, Henry completed *Ray* with *Charles*, not because he knew that Ray Charles was a famous individual, but because he had formed an unconscious associative link between Ray and Charles when he had heard or read those two names together. *Charles* just popped into his head. Henry completed fifty-one percent of first names with the last names of individuals who were famous before his operation, and a surprising thirty-four percent with the last names of individuals who were famous after his operation. He completed *Sophia* with *Loren*, *Billie Jean* with *King*, and *Martin Luther* with *King*. Although Henry's ability to acquire this kind of information was clearly impaired relative to healthy participants, the results suggested that he had at least thin knowledge of postoperatively famous individuals, sufficient to support a link between their first and last names.[28]

The next day we presented meaningful cues for each individual—"famous artist, born in Spain in 1881, formulated cubism, works include Guernica." After this cue, Henry heard, "When I say Pablo, what is the first word that comes to mind?" This kind of cueing boosted Henry's last name identification equally for individuals who came to prominence before and after his operation. The fact that he benefited as much from semantic cueing about post-1953 names as about pre-1953 names suggested that his new knowledge, like his preoperative knowledge, was incorporated into *schemata*. These organized semantic networks, clusters of related information, are capable of supporting conscious recall. This finding adds to the body of evidence that Henry was capable of limited declarative, semantic learning.[29]

In a critical companion experiment, we examined the scope of Henry's new semantic knowledge by focusing on the amount of detail he could provide about celebrities. First, he saw two names side by side. One was famous and one was a name randomly taken from the Boston area telephone directory. When asked, "Which is the name of the famous person?" he was correct ninety-two percent of the time for names he had come across before his operation and an impressive eighty-eight percent for those he had encountered afterward. Then, for each individual Henry selected as famous, we asked him a key question: "Why was that person famous?" Even though the semantic information Henry provided for post-1953 celebrities was impoverished compared with the responses of control participants, and compared with the information he himself furnished for individuals who were celebrities before 1953, the result was astonishing. He was capable of providing accurate, distinguishing information for twelve people who became famous after 1953. He knew that Julie Andrews was "famous for her singing, on Broadway," that Lee Harvey Oswald "assassinated the president," and that Mikhail Gorbachev was "famous for making speeches, head of the Russian parliament."[30]

This investigation demonstrated that some semantic knowledge can be acquired in the absence of any discernible hippocampal function. Henry, despite a lesioned hippocampus, showed that he was capable of learning information about celebrities who came to fame after his operation, providing robust, unambiguous proof that at least some semantic learning can occur in the absence of episodic learning.[31]

While it is interesting to note the extent to which semantic learning can occur without hippocampal function, it is equally interesting to note the ways in which this learning was different from that of our control participants. Henry was capable of generating semantic knowledge about only a fraction of the individuals who were well known to controls. Moreover, the information he generated about the postoperative people familiar to him was relatively sparse compared with control participants and compared with the amount of information he was able to generate about preoperatively famous individuals. Henry, for example, could not provide even the sex of some of the names he selected as famous. For instance, he said that Yoko Ono was "an important man in Japan." Further, while the control participants were better at generating knowledge about individuals who became famous recently compared with remotely, consistent with the general pattern of forgetting typically observed in healthy people, Henry showed the reverse pattern of performance. In addition, his ability to produce information about postoperatively famous individuals was sporadic. For example, in prior attempts to assess his learning about famous individuals, he had successfully identified Ronald Reagan as a president and Margaret Thatcher as a British politician, but was unable to generate their occupations during this investigation. And during the 2002 investigation, he indicated that John F. Kennedy had been assassinated, whereas on prior occasions, he had said that Kennedy was still alive.[32]

The restricted nature of the semantic knowledge that Henry was able to demonstrate makes it unlikely that the mechanisms he relied on for learning were identical to the mechanisms healthy adults use to acquire semantic knowledge so prolifically and spontaneously. Specifically, his ability to show any rapid learning of semantic knowledge was eliminated, presumably due to his bilateral hippocampal lesion. His only mechanism for learning was via *slow-learning*, in which extended repetitions of information enabled him to glean some information.[33]

In interpreting the results of this study, it was important to consider whether Henry's acquisition of limited semantic information did, in fact, represent declarative learning and not nondeclarative perceptual memory—information acquired automatically through visual exposure. Henry's learning differed from nondeclarative memory in several important respects. *First*, a hallmark of declarative memory is that it is accessible to conscious awareness and can be voluntarily brought to mind in words or images. In contrast, nondeclarative learning is accessible only through reenacting the task in which the knowledge was learned. Henry was able to freely recall specific details about a limited number of postoperatively famous people—John Glenn as "the first rocketeer"—or events—the assassination of John F. Kennedy. *Second*, the expression of nondeclarative memory is rigidly determined by the manner in which it was acquired, whereas semantic knowledge can be recapitulated flexibly in response to a variety of relevant stimuli. Henry repeatedly retrieved information about a small number of celebrities regardless of the specific language used to frame the question or the modality of the stimuli (words versus pictures). *Third*, Henry's ability to generate familiar last names when he heard the first names could have been explained as an automatic stimulus-response reaction supported by nondeclarative memory. But the fact that he benefitted as much from semantic cueing about postoperative names as about premorbid names demonstrated that this new knowledge, like his preoperative knowledge, had been incorporated into a semantic network capable of supporting conscious recall. Based on this evidence, we concluded that Henry was capable of meager declarative, semantic learning. Where in Henry's brain did this unusual learning take place? The likely candidates were the remaining bits of memory-related cortex near his lesion—perirhinal and parahippocampal cortex—and the vast cortical networks where information was stored.[34]

Why did Henry exhibit semantic learning about famous personalities in this study, but fail to learn new vocabulary in an earlier one? One possibility was the difference in the number and types of exposure to the stimuli. Celebrities offered a wider array of opportunities for encoding information, and Henry may have encountered the names John F. Kennedy and John Glenn on numerous occasions and in different, rich contexts. He watched the news every night from 6:00 to 7:00 p.m. and often looked at and read magazines. This variety of exposures in his everyday life may have given rise to richer and more flexible memory traces than processing the isolated words in the lab—*minatory, egress,* and *welkin.* Another possibility is that the names presented to Henry may have allowed him to take advantage, at least in some instances, of knowledge related to them that he learned before his operation. For instance, his ability to remember details about John F. Kennedy may have resulted from his knowledge of the Kennedy family acquired during the 1930s and 1940s. Similarly, Liza Minnelli had two famous parents, singer and actress, Judy Garland, and film director, Vincente Minnelli.[35]

Prior knowledge seemed to have helped Henry in a different kind of experiment, this one using crossword puzzles—Henry's favorite pastime. Our experiments carried out in 1998–2000 sought answers to three questions: How proficient was Henry in solving crossword puzzles compared to healthy participants? Could he solve preoperative clues linked to postoperative events? Would his accuracy or speed increase after repeated exposure to the same puzzles? Using test materials specifically designed for Henry, we gathered additional evidence that he could anchor new semantic information to old semantic memories. We created three kinds of crosswords of twenty clues each, incorporating semantic knowledge from different time periods. One used historical figures and events known prior to 1953 with clues such as: *baseball player in the 1930s who captured the home run record.* We called this the *pre-pre puzzle* and expected that Henry would be able to solve these clues. Another had clues based on historical figures and events popularized after 1953: *husband of Jackie Onassis, assassinated while president of the U.S.* We called this the *post-post puzzle* and expected that Henry would not solve these clues. The third puzzle combined the previous two time periods by giving post-1953 semantic clues for pre-1953 answers: *childhood disease successfully treated by Salk vaccine* (post-1953 knowledge); answer: *polio* (pre-1953 knowledge). We called this the *pre-post puzzle* and thought that Henry had a good chance of putting his old knowledge to work during the course of solving it. Henry's instructions were to complete each puzzle in any manner he liked, and he was allowed to erase answers. We did not impose a time limit but asked him to tell us when he had finished each one. Henry completed the same three puzzles once a day for six consecutive days. Each puzzle was presented only once on a given day, and he had a short break before beginning the next puzzle. At the end of each test session, the examiner showed him the correct answers. Henry corrected any misspelled words and inserted the correct answers where he had left blanks.[36]

We were curious to see whether the repeated exposures to the correct words on the third puzzle—combining pre-1953 clues with post-1953 answers—would engage Henry's semantic network of pre-1953 knowledge to the point where he would eventually fill in the correct answers. We believed this was a real possibility because our previous evidence showed that occasionally he could engage his existing mental schemas to acquire new facts (JFK was assassinated). On the pre-1953 puzzle, Henry responded with high accuracy and performed consistently well. But he regularly missed the two most difficult clues, *Chaplin* and *Gershwin*, and his overall score did not improve across the six days of testing. On the post-1953 puzzle, he was, not surprisingly, very inaccurate, and again his performance did not improve over six days. In marked contrast to the absence of learning on the pre-pre and post-post puzzles, however, Henry *did* show improvement across five days of testing with the pre-post puzzle because he could link the new information to mental representations established before his operation. He successfully learned to associate the postoperative knowledge with preoperative information for six answers: *polio, Hiss, Gone with the Wind, Ike, St. Louis,* and *Warsaw.* This improvement is consistent with the general idea that new semantic learning in amnesia is facilitated when the information is meaningful to the individual patient, something that the person can relate to.[37]

In the crossword puzzle experiment, Henry demonstrated an ability to learn the solutions to puzzle clues that allowed him to benefit from preoperative knowledge. This same mechanism—linking postoperative to preoperative knowledge—was at work when he told us facts about some of the celebrities whose names he recognized as famous. He was able to encode, consolidate, store, and retrieve a small amount of information about famous people—knowing that John F. Kennedy "became president; somebody shot him, and he didn't survive; he was Catholic."[38]

The concept of *mental schemas* sheds an interesting light on Henry's unexpected ability to consolidate and retrieve the occasional piece of new semantic knowledge. Sir Frederic Bartlett, a British philosopher who became an eminent experimental psychologist, introduced the concept of schemas in 1932. Based on his studies of memory performance in healthy research participants, Bartlett wrote, "Remembering is not the re-excitation of innumerable, fixed lifeless and fragmentary traces." Instead, he viewed remembering as an active process—the capacity to creatively rebuild your inner representations of the world. He named these organized, constantly changing masses "schemata." As Henry was trying to solve the pre-post crossword puzzles, he may have been relying on an enduring structured representation of old knowledge—a schema—to understand, store, and recall the new information.[39]

When we watch a political debate, we see the candidates presenting details about their policies and how they will implement them. As the questions and answers unfold, we feed the new information into a mental framework that allows us to understand, evaluate, and consolidate each candidate's ideas. Some time later, prior to election day, we can consult our updated mental schema and make an informed decision about whom to vote for. We are able to make our choices efficiently because we have stored the relevant semantic information in an organized body of knowledge. Henry retained mental schemas established during the years before his operation, and occasionally he was able to tap into them to anchor a few new facts.

In 2007, neuroscientists at the University of Edinburgh conducted experiments on schema learning in animals. They trained normal rats to associate distinct food flavors with specific locations in a small arena that was familiar to them. Initially, the rats formed six flavor-place associations. By trial and error, they learned, for example, that rum-flavored food pellets were in one location, banana-flavored pellets in another, and bacon-flavored pellets in another. There were six sand wells where the rats could dig for their reward. During learning, the animals were cued with a specific food in the start box, and their task was to find the well that contained the same food (cued recall). If they dug in the correct sand well, they were rewarded with more of the cued food. After several weeks of training, the animals acquired associative schemas for this task—they mapped each flavor to a specific sand well in the arena.[40]

The researchers then asked whether having this schema would facilitate the encoding and consolidation of new flavor-place associations and their rapid integration into the existing schema. They closed two of the sand wells and introduced two new ones with two new flavors. The rats received just one rewarded trial with each of the two new pairs and then rested for twenty-four hours. When the researchers tested their memory for the two new flavor-place associations, the rats chose to dig at the correct locations and not at the closed ones. They learned the new pairs in a single trial and remembered them for twenty-four hours, indicating that the earlier learning of the associative schema helped in this process. After another twenty-four hours, the researchers gave the rats hippocampal lesions. When they recovered from the operation, the rats still recalled the location of the original schema and, amazingly, the two new pairs. The new associates had been quickly consolidated and stored outside the hippocampus, probably in the cortex. Apparently the rats had learned an associative schema incorporating the mapping of flavors to places in the arena, and this schema provided a framework to help them retain the two novel paired associates.[41]

During the twenty-seven years before his operation, Henry had successfully built up numerous schemas and stored them in his cortex. Even though his medial temporal-lobe lesion prevented him from learning new associations, like *cabbage-pen*, he could sometimes fall back on a reservoir of schemas he had consolidated before his operation and had retained in his long-term memory. For instance, when solving the pre-post crossword puzzle, he correctly responded with *polio, Hiss, Gone with the Wind, Ike, St. Louis,* and *Warsaw,* and this learning may have relied on his preoperatively acquired schemas. This kind of organized, stored information may be what enabled him to consolidate a few new facts after his operation. What he saw and heard on television may have activated and updated longstanding schemas related to politicians, movie stars, and technology, enabling him to remember JFK, Julie Andrews, Lee Harvey Oswald, Mikhail Gorbachev, and to define Skylab as "a docking place in space."

How did Henry's capacity to learn fragments of general knowledge affect his everyday life? My guess is that his sense that he knew a few people at Bickford, and his ability to recognize a name here and there, gave him the feeling that he was among friends. In 1983, when he returned to Bickford from MIT, a staff member noted that he appeared happy to be back and seemed to remember his companions. When he watched television, some of the news anchors and actors in sitcoms must have looked and sounded familiar, so he could relate to them as his TV buddies. Henry acquired factual knowledge about his nursing home: the layout of his room, the lounge, and the dining room, the dog who sat by his wheelchair, the woman who flirted with him, and the numerous aides who cared for him. Although his interactions with the world were far from normal, his life did have familiar attachments that helped him feel secure. Overall, in spite of his tragedy, Henry got along.

We have numerous, attention-grabbing examples of Henry's ability to acquire snippets of novel semantic knowledge after his operation. Still, his consistent impairment compared with controls made it clear that the surgical removal of his medial temporal-lobe structures in 1953 had decimated his ability to acquire a *significant* amount of new semantic information. In spite of this gap in his knowledge, however, he was capable of thinking about his personal world and communicating effectively. He had an excellent vocabulary and an impressive knowledge of world events and celebrities, but this knowledge was frozen in time, an archive of information from the first half of the twentieth century.

Twelve

Rising Fame and Declining Health

After the publication of Scoville and Milner's 1957 paper "Loss of Recent Memory after Bilateral Hippocampal Lesions," Henry gradually became famous within the neuroscience community. His story began to appear in psychology and neuroscience textbooks in 1970, and by the 1990s was invoked as a case study in nearly every textbook that addressed memory. In scientific papers, he was frequently highlighted as the inspiration for particular experiments. Every young psychologist and neuroscientist learned about H.M. in school, and the description of his amnesia was a touchstone for the severity of memory impairment in other patients. Through our continued research with him, Henry became the most comprehensively studied patient in neuroscience.[1]

By the late 1970s, I had become Henry's primary point of contact for anyone who wanted access to him for research. Hans-Lukas Teuber died in 1977, and Brenda Milner moved on to other research topics, while still maintaining a strong interest in Henry. I had inherited him as a patient. He lived only two hours away from MIT, so it was logistically easy for him to visit my lab or, as he grew older, for my colleagues and me to visit him at Bickford.

265

Over the years, a number of researchers came to MIT to test Henry for their own studies, but I felt strongly that Henry should not be made available to every person who wanted to meet him. If I had opened the gate, allowing all interested researchers to test and interview him, the resulting free-for-all would have been a constant drain on his time and energy, and would have taken unfair advantage of his memory impairment and willingness to be helpful. Many people were eager simply to talk with Henry, but I did not want him to become a sideshow attraction—the man without a memory. I, therefore, required any investigator who wanted to study Henry to visit my lab first and present the proposed research protocol at our weekly meeting. I wanted to make sure that the experiments were well designed, so that the data from Henry would generate meaningful conclusions. My requirements may have frustrated some, but they spared Henry from being besieged by frivolous inquiries.

From 1966 onward, 122 scientists had the opportunity to work with Henry, either as members of my lab or as our collaborators from other institutions. Together, we investigated a broad range of topics. A memory researcher from the University of California, San Diego, came to study Henry's semantic knowledge. A vision scientist at the Rowland Institute in Cambridge, Massachusetts, examined an aspect of visual perception in Henry and a group of Alzheimer patients to find out whether memory impairment affected performance on a visual aftereffects task. A neuroscientist came from the University of California, Los Angeles, to conduct EEG recordings while Henry detected different targets on a screen.

Although every visiting scientist had read extensively about Henry and his case, some still found the experience of meeting him in person astounding. One colleague, Richard Morris, recalled meeting Henry with a group of hippocampus researchers. Later, he wrote this note to me describing the event:

> We sat in a room, and he came along, and we met him. In many respects, he was exactly as you had described him in papers— very courteous, very polite. Initially, the conversation was such that you had no reason to think there was anything at all

untoward. This was just like meeting a very kindly, genial old man, really. But then gradually one or two things happened, various repetitions of things that began to reveal that there was something untoward.

The opportunity arose for one of us to leave the room, and it actually was I. We'd been talking maybe for a half an hour or so, and then I got up and left the room and deliberately stayed out of the room for about ten minutes and then rejoined the conversation. It was very striking because my colleagues introduced me to him again and he said, "Nice to meet you," as if he hadn't known that I'd been there before, and then pointed to an empty chair and said, "There's an empty chair. You go sit there." And this, of course, had been the chair I'd been sitting in before. So that was exactly as we'd been led to expect from the published literature, but to see it with our own eyes was interesting.

Henry lived a private and highly circumscribed life. Members of my lab and other colleagues who interacted with him, as well as the staff at the MIT CRC and Bickford, were all extremely careful to keep his real identity secret; until his death, he was known to the outside world only as H.M. Over the course of twenty-five years, however, he and his story became increasingly famous. His case intrigued journalists, artists, and the general public, while raising ethical questions about experimental medical interventions. People were fascinated by the story of such profound and ongoing amnesia.

The staff at Bickford knew that Henry had a special importance to the world, and I instructed them not to discuss him and his case outside the nursing home. I received many requests from media outlets to interview and videotape Henry, but I shielded him from too much exposure, and asked my lab members not to take photographs or videos of him. To my knowledge, no videos of him exist. I did allow a science writer, Philip J. Hilts, to meet Henry and spend time in my lab for his 1995 book *Memory's Ghost: The Strange Tale of Mr. M. and the Nature of*

Memory. Hilts talked extensively with Henry and even had his own desk in my lab while he was gathering information.

Excerpts of my 1992 interview with Henry that appeared in subsequent radio programs helped the public, as well as scientists, put a human voice to the more impersonal language of Henry's case study. This conversation also revealed his memory deficit unmistakably—he repeated the same thoughts several times and was unable to tell me the current month and year, or what he had for lunch.

I was with Henry when he was admitted to Bickford in December 1980. Mrs. Herrick drove him and his few belongings the fifteen miles from her home in Hartford. He was assigned a room on the second floor of the building, a small facility decorated in pastel greens, flowered wallpaper, and oak furniture. Henry adjusted well to his new life at Bickford, and the staff described him as mild-mannered, cooperative, and compatible with the other patients. After Mrs. Herrick died the following year, the nursing home became the center of Henry's universe; he spent the final twenty-eight years of his life as a resident there.

At Bickford, Henry had a new lifestyle, one that gave him the constant care of staff members, along with a more social group environment. Over nearly three decades, Henry lived in different rooms (often with a roommate), saw the facility undergo extensive remodeling, and interacted with many different nurses and aides, some of whom stayed there as long as he did. During that time, Henry became a beloved fixture at the home, well known and liked by everyone. Members of my lab sent him pictures of themselves so that his bulletin board would have the same human touch as other patients' rooms.

For Henry, each encounter was fleeting because of his inability to store memories of events and facts, but although he could not remember names or details, many staff members believed that he had a sense of who they were. Henry was an unusual patient in this setting. At fifty-five, he was younger than most of the other residents when he arrived. In addition, he was intelligent, alert, and in relatively good health. In some ways, however, caring for Henry was similar to managing the patients with dementia. They also forgot recent events and information, such as the nurses' names, while they could still talk about their childhood and the town where they grew up. Like the demented patients, Henry needed prompting to carry out even the simplest daily activities. He was a mentally alert adult man who nevertheless had to be supervised and directed like a child.

Henry was always courteous toward the staff at Bickford. He greeted them with his wide, dimpled smile and apologized frequently for needing their help. For someone with such a severe memory problem, Henry was surprisingly easygoing. He was cheerful and never seemed uncomfortable or nervous when interacting with nurses and aides, behaving as if everyone were an old friend.

Unlike the simple life he had spent at home with his parents and then with Mrs. Herrick, Henry's time in the nursing home offered many opportunities for engaging in activities and interacting with lots of people. Overall, he was a sociable person one-on-one but a quiet participant in group situations. He willingly took part in all sorts of activities—choral practice, Bingo, Bible study, film watching, poetry readings, arts and crafts, and bowling. He spent time in the lounge watching TV and liked to sit in the small courtyard in front of the facility. He continued to do crossword and other word puzzles; my lab gave him a monthly subscription to a crossword publication, so he always had a fresh supply. He attended special events held at the home and even danced the hula at a Hawaiian party. With his crafts projects, he was a perfectionist, displaying the attention to detail he had shown in the skilled labor of his younger days. One of my treasured possessions is a wooden spoon that Henry had painted blue and decorated with sponge-texture and small flowers with white petals and red centers. The Bickford staff kindly gave it to me the last time I visited him.

Henry was an animal lover. I have a touching picture of him as a young teenager, cradling two kittens in his arms. Fortuitously, Bickford was home to a variety of pets. For a while, they had a rabbit that Henry liked to hold in his lap. A singing cockatiel, Luigi, was in residence for more than ten years; other inhabitants included lovebirds, finches, and parakeets. Henry enjoyed watching the birds. A black-and-white dog named Sadie arrived as a puppy and remained there throughout Henry's stay. In the last years of his life, she often sat next to his wheelchair while he patted her, and she attended his burial.

Although Henry could not form relationships in the usual sense, he did interact with other patients and had an easy rapport with the staff at Bickford. He ate his meals with fellow residents at round wooden tables in the dining room, and he seemed to prefer the company of men. A couple of years into his stay, he made friends with another patient, Charlie, and they often watched TV together. In 1985, staff members reported that Henry enjoyed poker and woodworking groups designed for men only. He told the nurses that he liked to "let his hair down" and be one of the guys. Later in life, he did become friends with a female patient named Peggy, and the two were once crowned Prom King and Queen of the nursing home.

Henry never showed a romantic interest in women, however, and always behaved with propriety. In fact, he once had trouble with an attractive female patient who made sexual advances toward him in public, which embarrassed and confused him. The nurses told me that he would spend time with her only if she avoided using "sexually inappropriate suggestions and language." At these times, Henry would respond, "Oh, I can't do that. My doctor told me I can't do that." But if she was well behaved, "he would hold hands with her, and they would talk and talk about everything, and it was really nice."

Life in the nursing home, however, was not always smooth for Henry. His mood was sometimes marked by confusion, frustration, and anger. From time to time, he would become agitated if another patient was loud or wanted to watch a TV show that he did not. He could become irritated easily by noise or other physical discomforts, such as abdominal and joint pain. Although he continued to be a personable, good-humored patient, he sometimes appeared anxious and exhibited bizarre behavior. Much of it undoubtedly arose from his memory impairment and his difficulty communicating with the staff about his needs and wishes. Identifying and fixing the source of his problems was complicated. Henry could become confused and unable to communicate the exact reason for his irritation; instead, he acted out. When he had trouble dealing with anger, pain, sadness, or frustration, he might respond by hitting something, throwing an object, or threatening to jump out the first-floor window.

One night in 1982, Henry left his bed and stumbled out of his room, yelling that other patients were making noise and keeping him awake. He swung at staff members and slammed his hand against the wall, missing a nurse. Two police officers were called to the scene; Henry was given anti-anxiety medication, and began to calm down. The next day, when asked if he remembered what had happened the night before, he answered, "I don't remember—that's my problem." When pressed whether he remembered two big policemen, he said, "Sometimes it's better not to remember."

During the autumn of 1982, Henry became angry more often than usual. His displays of strong emotion were reminiscent of his outbursts in 1970 at home with his mother and at the Regional Center where he worked. Henry never exhibited this behavior when he visited us at the MIT CRC, where he had his own room and the environment was more tranquil. There, he received VIP attention and had lots of positive social interactions. His visits with people from my lab kept him in good spirits, bringing out the best side of his personality; he enjoyed the mental and social stimulation. Henry's emotional episodes and irritability at Bickford forced the staff in October 1982 to consider moving him to a nearby mental health care facility in Newington, Connecticut. Luckily, and for no apparent reason, his behavior returned to normal, so a transfer was unnecessary. After that episode, Henry sometimes went through bad times, but they were never more than the staff could handle.

Beginning in the 1980s, Henry consistently identified me as a friend from high school. The most likely explanation for this false recognition is that during his multiple visits to MIT, which took place on a regular basis from 1966 through 2002, he steadily built up a mental schema incorporating and linking my face, name, profession, and the CRC. He must have felt a vague sense of familiarity for me. If I gave him a list of last names all beginning with *C* and asked which one was my name, he chose *Corkin*. In 1984, I posed the question, "Who am I?" and Henry replied, "Doctress . . . Corkrin." (He used that unique Henryism, "Doctress," on several occasions.) When asked where he knew me from, he replied, "East Hartford High School." I had the following exchange with Henry during our 1992 conversation:

SC: Have we ever met before, you and I?

HM: Yes, I think we have.

SC: Where?

HM: Well, in high school.

SC: In high school.

HM: Yes.

SC: What high school?

HM: In East Hartford.

SC: And what year was that—about?

HM: 1945.

SC: Have we ever met any other place? . . .

HM: Or '46. See, I come way back myself in a way, because I remember I skipped a year in school.

SC: Right. Have we ever met anyplace besides high school?

HM (*pause*): Tell you the truth, I can't—no. I don't think so.

SC: Why am I here now?

HM: Well, you're just having an interview with me, I say. That's what I think right now.

During the interview, I neglected to ask him what my name was, so I asked him on the way back to his room. He first said, "I don't know," but then said, "I think of Beverly." I said, "No, it's Suzanne." Then he said, "Suzanne Corkin." In May 2005, a nurse at Bickford said to Henry, "I was just talking to a friend of yours in Boston, Suzanne," and he immediately said "Oh, Corkin." Although he had formed an association between my first and last names, when asked *explicitly* who I was, he did not know.

One name, Scoville, did stick in his mind because he had learned it before his operation. At Bickford, Henry often referred to Dr. Scoville, and he linked the name to the sense he had of his own importance to medicine. He even exploited the connection to try to get his way, telling people that "Dr. Scoville said you should do this," or "Dr. Scoville says it's supposed to be this way." Once, Henry's roommate got fed up and said he was tired of hearing about "Dr. Screwball." Henry quickly corrected his roommate's pronunciation.

In addition to his memory impairment, Henry began to deal with the degradations of old age, developing an increasing number of physical complaints and struggling with simple daily tasks. He had numerous falls, some followed by unconsciousness or injury; in 1985, a tumble resulted in the fracture of his right ankle and left hip. The following year, at age sixty, he had his left hip joint replaced. After both incidents, he went through several weeks of physical therapy until he could use a walker on his own. Henry had a history of difficulties with locomotion, and always walked with a slow, awkward gait. Even so, he was able to master the walker—a testament to his ability to learn new motor skills—but he occasionally forgot he needed the walker and tried unsuccessfully to walk on his own. On a few occasions, these attempts resulted in more falls. The contrast between the integrity of his nondeclarative memory and his lack of declarative memory was striking—and in this case dangerous.

During his hip-replacement hospitalization in July 1986, he had a grand mal seizure and a transient fever. His doctor wrote to me that after the surgery Henry was "troubled by insomnia, nocturnal anxiety, and fear of being alone." He would complain of an upset stomach and buzzing in his right ear, as well as pain in his hip area. It took him some time to recover fully from the surgery. In September 1986, when he visited the MIT CRC, the nursing staff noted that Henry was behaving strangely; he rang his bell frequently with complaints, slept restlessly, made nonsensical comments, and had a "wild-eyed expression." One nurse noted that Henry was not smiling and telling his usual stories and jokes. This temporary alteration in Henry's behavior was probably a side effect of the anesthesia administered during his hip replacement. Older individuals do not process anesthetics as well as young adults, and the residual substances may have interacted with pain medications given to Henry after his operation. Anesthesia side effects may last three months or longer in older adults because of the longer time needed for the drugs to clear out of their bodies. Eventually, the odd behavior stopped and Henry returned to normal.

Henry still had occasional seizures, but major attacks came infrequently—only once or twice a year—and some years were seizure-free. He saw a local doctor every other month and occasionally went to Saint Francis Hospital in Hartford for emergency care. Sometimes, during visits to MIT, he consulted with and had procedures performed by doctors in Cambridge. Henry's memory loss made it impossible for him to give an accurate account of his physical complaints, creating an extra challenge for the doctors who evaluated and treated him. A neurologist at the Mass General who examined Henry in 1984 noted that he tended to downplay the significance of his medical problems. He sometimes worried about being a bother to people, and this attitude may have led him to gloss over his physical ailments.

One symptom that regularly plagued Henry was his tinnitus, a ringing in the ears, which is a common side effect of Dilantin. He gave the neurologist the gist of his tinnitus but could not recount specific details about when it occurred or got worse. One nurse at Bickford gave us a fuller account, however, describing the devastating recurring episodes Henry had suffered for three to four months in 1984. At first, they occurred several times a week, and then once a day. They began early in the morning when Henry was still in bed. The nurses often found him with a pillow over his head, irritable and refusing to be touched or helped. On several occasions, he asked for a gun so he could kill himself to end the suffering. During these episodes, he reported a high-pitched, shrill sound that persisted for two to eight hours. They were not seizures, and an examination by an ENT doctor did not uncover a trigger in his inner ear. The Mass General neurologist instructed Bickford to substitute Tegretol for Dilantin to control Henry's seizures, and the torturous episodes decreased.

Although subdued, Henry's tinnitus persisted and contributed to his bouts of discomfort and agitation. He had a particular sensitivity to noise, complaining often about the racket from other patients and the air conditioner in his room. The Bickford staff found him several times putting cotton balls in his ears to relieve the ringing, and when it was particularly bad, he would refuse meals—a noteworthy occurrence, given Henry's healthy appetite. In addition to his tinnitus, Henry complained of vague symptoms such as stomach pain or neck discomfort. The staff found that simply asking him to identify the problem was not helpful because he had already forgotten the source; instead, they had to ask him specific yes-or-no questions to find a remedy. Sometimes he would stay in bed, and the discomfort would resolve itself.

Henry was part of my life for decades, and although I had to maintain my unbiased role as a researcher, it was impossible not to care about this pleasant, gentle man. In 1986, when Henry celebrated his sixtieth birthday at MIT, members of my lab and the CRC staff organized a party in his honor; our rendition of "Happy Birthday" brought a big smile to his face—as did the cake and ice cream. We always tried to make him feel like a member of our team by remembering his birthday, sending him Christmas presents, and making sure he always had his crosswords. Toward the end of his life, I contacted the staff at Bickford on a weekly basis to check up on him.

Throughout the 1990s, Henry had good and bad days at the nursing home. On good days, he greeted the aides with his characteristic wink and smile; on bad days, he moaned about pain or discomfort. His usual pace became more sluggish, he spoke and moved slowly, and he needed help with everyday tasks. He became less willing to move about on his own, sometimes traveling to the dining room for his meals in a wheelchair. In 1999, a fall led to another broken ankle, which further hampered his independence. Still, he remained relatively well through 2001, after which he was beset by further maladies. He developed osteoporosis, suffered from sleep apnea, and had intermittent high blood pressure. After he died, we learned from his autopsy report that, despite regular medical care, he had been afflicted with several undiagnosed disorders: atherosclerosis, kidney disease, and colon cancer. One or more of these conditions may have been the source of his incontinence and constant need to use the bathroom as he got older. At age seventy-five, he became fully dependent on a wheelchair to move from place to place. Even then, however, his intellect and sense of humor remained intact. In March 2002, during a visit to MIT, a researcher asked Henry whether he had slept well. "I didn't stay awake to find out," he replied.

From 2002 through 2004, Henry was still healthy enough mentally and physically to take part in formal cognitive testing. Because of his motor difficulties, it became uncomfortable for him to travel to MIT, so we drove to Bickford to carry out our behavioral experiments. During this time, however, Henry did make several trips to the Mass General Martinos Center for Biomedical Imaging for a series of MRI scans. He traveled by ambulance to ensure that he would have a comfortable ride. Each time he came for one day, and we hired two aides to care for him. As a result of cutting-edge advances in MRI, these studies gave us a clearer picture of Henry's lesions and new insights about his aging brain. Researchers used the MRI images to create a computerized model of his brain that could be compared with those of healthy male participants of the same age. The scientific question was whether the changes in Henry's brain were equivalent to or exceeded those in the controls. A related question that emerged as Henry aged was whether his brain would show evidence of aging-related disease, independent of the damage caused by his 1953 operation. Having this information would give us a complete picture of his neural infrastructure and help us understand his cognitive capacities late in life.[2]

Even to the naked eye, a normal eighty-year-old brain looks dramatically different from a normal twenty-year-old brain. The total volume is reduced in the older brain, and consequently the fluid-filled spaces in the middle of the brain—the ventricles—become larger. Because of this tissue shrinkage, the valleys—sulci—are deeper, and the peaks—gyri— are thinner, thus accentuating the folds in the cortex, the outer surface of the brain. The loss of mass is not uniform, however; some areas look markedly smaller in old age, while others remain relatively unchanged. With modern MRI technology, we can spot these changes in healthy participants over as short a time as one year. The brain atrophy observed in healthy older adults told us what to expect in Henry's brain, simply as a consequence of natural aging processes. Research in my lab also alerted us to the importance of white matter integrity for the performance of complex tasks.[3]

Some scientific questions linger: how do these physical changes in the brain correspond to changes in specific cognitive abilities, and which of the identified late-life alterations best explain cognitive loss? Even when people age successfully, they show some cognitive decline. Hardest hit are working memory—for example, mentally dividing a large dinner bill by fourteen to determine the amount that each guest owes—and long-term memory—remembering the names and identities of all the people encountered at a wedding. All complex processes in the brain, such as memory and cognitive control, depend on interactions within and among specific brain regions. Because white-matter tracts, the lines of communication, must be kept open and running smoothly for optimal function, it is reasonable to expect that damage to white matter in a particular region would impair the cognitive capacities that depended on that region.[4]

Because I kept in touch with Henry and had access to his nursing home records, I was aware of the ups and downs during this last third of his life. By 2000, his gradual decline had continued. He now had some difficulty or perhaps unwillingness to feed himself, even though he continued to gain weight. He had few seizures but showed some cognitive impairment—for example, decreased concentration, lessened ability to process instructions, and increased confusion. An additional loss of coordination and strength made it challenging for Henry to travel around the nursing home, and he continued to experience falls. His social life also went downhill; he often chose not to participate in group activities and had trouble making friends, instead showing frequent periods of irritability, anger, and obsessive behaviors, such as a preoccupation with going to the bathroom. He had a tendency to worry and become impatient; as he observed, "I'm always on edge." His speech was slurred, perhaps from too much medication, and he slept a lot in his geriatric chair. From time to time, Henry's oxygen level was low, so the staff gave him oxygen via tubes in his nostrils. Usually, he took the tubes out, and once put them in his urinal.

In 2005, Henry had several grand mal seizures, his cognitive and motor capacities declined further, and he became completely dependent on others. In spite of these handicaps, his daily life seemed to have improved; he attended activities three to five times a week, and according to his chart was "very social with his peers." He still enjoyed Bingo, crossword puzzles, and word games, and one note described him as a "pleasant interesting person—a joy to talk with." It is safe to say, however, that by this time Henry was demented, based on the global deterioration of his cognitive capacities.

What is the difference between amnesia and dementia? Pure amnesia, such as Henry experienced after his operation, consists of memory impairment without additional cognitive deficits. In contrast, dementia is characterized by severe memory loss compounded by defects in multiple cognitive domains—language, problem solving, math, and spatial abilities. Dementia goes far beyond the changes that accompany healthy aging, and by 2005, Henry had crossed over from amnesia plus healthy aging to amnesia plus dementia. We found its neural underpinnings in his MRI scans.

The imaging studies we conducted from 2002 to 2004 gave us a more complete picture of Henry's aging brain and complemented our years of rich clinical observations. The effects of his operation were compounded both by changes related to his advancing age and by new brain abnormalties. The MRI scans showed small strokes in his gray and white matter that were not connected to his operation, but that stemmed from aging-related disease—probably advanced white-matter disease caused by hypertension. These localized regions of dead brain tissue caused by a lack of blood and oxygen were in the expected areas for brain disease from high blood pressure. We also saw small strokes in gray-matter structures under the cortex, such as the thalamus, an area that integrates sensory and motor activities, and the putamen, a motor area under the frontal lobes. An accumulation of such strokes was the likely cause of Henry's dementia. A further correlate of his dementia came to light when we compared the thickness of his cortex to that of control participants. We found that thinning had occurred to a greater extent than normally expected for his age and was widespread across his entire cortex, rather than focused in certain areas as we typically see with healthy aging.[5]

We had not seen most of these alterations in his 1992 and 1993 scans, suggesting that they had originated recently. We knew from our MRI studies in healthy older adults that their cognitive capacities are closely linked to the integrity of white-matter tracts in their brains: participants with the most intact white matter had the best memory-test scores. Knowing this, my lab focused on the status of Henry's white-matter communication system. We found widespread damage to Henry's white matter—more extensive and severe than normally expected from healthy aging. A further analysis based on diffusion-tensor imaging showed that the white-matter fibers had lost some of their structural, and presumably functional, integrity. He also had white-matter damage consequent to his operation.

We will pursue our examination of Henry's white matter by analyzing the MRI images of his autopsied brain. A new tool, tractography, will allow us to trace specific fiber bundles to localize white-matter damage. We will compare the results of this analysis with the anatomy derived from studying the actual white-matter tracts in his autopsied brain. Based on the characteristics of the fiber-bundle damage, Matthew Frosch,

Director of Neuropathology at Mass General, will be able to distinguish the specific white-matter pathways related to Henry's operation from those related to the strokes. We will know definitively what kind of dementia Henry had after Frosch conducts a detailed neuropathological examination of his brain tissue. Many lingering questions about Henry's brain must await upcoming microscopic analysis.[6]

During the 1980s, when I was studying Alzheimer disease, I learned the importance of brain donation: a definitive diagnosis of the disease can be made only at autopsy. As Henry began to decline, I focused on ensuring that we would be able to study his brain after he died. Although we could learn a great deal through brain imaging, the only way to learn conclusively about the status of the remaining tissue was to examine it microscopically. We would then know with certainty what gray matter and white matter had been removed during his operation and what was spared. Additionally, we could document any abnormalities related to advanced age and aging-related diseases. I explained to Henry and his court-appointed conservator, Mr. M. (Mrs. Herrick's son) the importance of examining Henry's brain after his death, and asked whether Henry would donate his brain to Mass General and MIT. They completed an Authorization for Brain Autopsy form in 1992.

In 2002, I assembled a team of neuroscientists for the first of many meetings to plan the details of what we would do step by step when Henry died. I selected colleagues who brought different kinds of expertise to the project in hopes that we could do something extraordinary. Henry had already contributed much to the world's knowledge of memory, and I wanted to extend the research by applying cutting-edge techniques for imaging, preserving, analyzing, and disseminating information about his brain. The team included neurologists, neuropathologists, radiologists, and systems neuroscientists from Mass General, the University of California, Los Angeles, and my lab at MIT. Our discussions, carried out over the next seven years, identified several key tasks and the order in which they had to be carried out.

We knew it would be vital to harvest Henry's brain as soon as possible after he died, before the tissue deteriorated. To reach this objective, we laid out a set of plans in collaboration with Bickford—having a nurse or doctor officially pronounce Henry as dead and noting the time of death, packing his head in ice to preserve his brain, and arranging for a funeral home to transport his body to the Martinos Center at Mass General for pre-autopsy scanning. We created a telephone call-list to notify all the researchers that Henry's body was on the way. Once there, it would be necessary to transfer it from the gurney it traveled on to a smaller nonmagnetic gurney and from there into the scanner, all the while adhering to the safety precautions required for dealing with human tissue and fluids. After this *in situ* scanning (imaging of the brain in the head), we would transport Henry's body to the Mass General Morgue so Frosch could remove Henry's brain from his skull. The neuropathology photographer would be on hand to catch the first photographs of Henry's physical brain. His body would then be moved to another area in the pathology department where a general autopsy would be performed. To preserve the brain, the neuropathologist would have to order the proper solution in advance, so he would be equipped when the time came. After much debate, we decided that we would scan Henry's brain again, after it had been preserved for ten weeks. We deliberated about using a three Tesla scanner or a seven Tesla scanner, and ultimately decided to use both. Pilot scans on other autopsied brains had established that the procedures were safe.

We also worked out the logistics of transporting a brain from Boston to San Diego, California, where it would eventually be frozen in gelatin and cut into ultra-thin sections for microscopic analysis. We would be right on the threshold of looking at the actual neurons inside the most famous brain in the world. The big payoff would come when experts in medial temporal lobe anatomy and neuropathology would stain and examine selected sections, yielding answers to questions about Henry's brain that had been waiting in the wings for decades.

Our goal during Henry's final years was to understand the abnormalities we documented in his brain in relation to his clinical condition. Could these anatomical changes account for his slipping mental status? I administered a few cognitive tests during my yearly visits, and each time invited along a colleague from the Mass General Neurology Department to update Henry's neurological condition. These annual evaluations uncovered a clear pattern of cognitive decline due to aging-related abnormalities in his brain—documented in his MRI scans—in concert with toxic side-effects from multiple psychoactive medications. Dehydration may also have played a role in his compromised mental status. I do not know how much water Henry drank or the extent to which other factors, such as medication side-effects, contributed to his dehydration, but given his characteristic silence about hunger and thirst, he likely did not ask for water. Based on our clinical observations during Henry's life, we could not specify with confidence the cause of his dementia. It could have been Alzheimer disease, vascular dementia, or a combination of abnormalities.

In June 2005, when Henry was seventy-nine, I drove down to see him with a neurologist who had often examined him at the MIT CRC and knew him well. At the time of our visit, Henry's blood pressure was high, and he weighed 218 pounds. The examination revealed that although Henry's speech was slightly slurred and difficult to understand, he named four out of five common objects (he missed *stethoscope*), performed five gestures on command (such as *salute*), and could mimic gestures. His muscle strength was reduced, especially in his legs—not a surprise considering that he had spent recent years in bed or in a wheelchair.

After the neurologist completed his exam, I stood next to Henry's wheelchair and administered a few cognitive tests. Encouragingly, his digit span—his immediate memory span—had not changed; he could still repeat five digits back to me. So when I said, "Seven, five, eight, three, six," he immediately responded, "Seven, five, eight, three, six." His performance indicated that when he was at his best, Henry could still pay attention, follow directions, and respond appropriately.

Bolstered by that success, I asked him to define a list of vocabulary words. Some of his definitions were concrete, a general characteristic of brain-damaged individuals; when I asked him the meaning of "winter," he said, "cold," and for "breakfast" he said, "eating." For other words, however, he gave excellent definitions, telling me that *consume* means "eat," *terminate* means "end," and *commence* means "start."

I next evaluated his ability to name common objects by showing him line drawings and asking him to tell me what they were. This task, often used in tracking disease progression in Alzheimer patients, measured his semantic knowledge, the meaning of words. Of the forty-two pictures, he named more than half correctly, but this score was well below those of healthy control participants. In some cases, his response was marginal. When I showed him a tennis racket, he said, "for tennis," and a toboggan, "bobsled." He clearly knew what these pictures depicted but could not access their correct names. I am sure that fatigue played a role in his impoverished performance because at one point I noticed that he was nodding off. Nevertheless, Henry's brain had clearly lost some semantic knowledge.

Henry's lethargy was surely in part a result of the medications he was taking. His doctor had prescribed Xanax, Seroquel, and Trazodone. According to the physician's order sheet, Henry received these medications for agitation, anxiety, obsession with his bowels and bladder, and depression. These psychiatric symptoms were not part of his amnesia but were instead connected to his progressive dementia and undiagnosed colon cancer. At this point in his life, Henry, sadly, was a pharmacy in a wheelchair.

By 2006, Henry was declining sharply. The neurologist found numerous changes in his condition since the previous year. His blood pressure was low, although it had been quite high the previous year. He was somnolent, performed only three out of five gestures, and had limited arm movement. His strength in his hands and legs had decreased further. We speculated that the decline in his condition was due to new small strokes, brain degeneration, heart disease that affected the blood supply to his brain, or sedating medications—or a combination of these factors. The neurologist recommended that Henry's doctor at Bickford reevaluate the need for Xanax, Seroquel, and Trazodone. Henry required full-time care, spending his days in bed or in a geriatric chair, which reclined and was more comfortable and supportive than a standard wheelchair. He could feed himself now and again but most of the time needed help from the staff. He sometimes participated in group activities, especially Bingo and a daily coffee hour at which the residents did physical exercises. Henry could be attentive but tired easily.

By 2007, Henry's lethargy and confusion had increased. During our three-hour examination, his level of alertness waxed and waned from fully alert and interactive to sleepy, with his eyes closing from time to time. He made good eye contact and had an excellent social smile. When he was wheeled into the examination room, he looked around with interest at his four visitors and smiled at everyone. When we asked him how he was, he said that his right knee hurt. Knowing the scarcity of his pain reports in the past, this new pain must have been extreme to cause him to complain. The physical exam showed that his knee was mildly swollen and warm, so the neurologist prescribed ibuprofen and recommended that Henry increase his intake of fluids to combat dehydration. We could see that he was dehydrated because when the physician gently pinched the skin on the back of his hand, it did not bounce back as it would in a well-hydrated person. He was still oversedated, so the physician also recommended decreasing the doses of several medications.

By this stage, Henry's language was fluent but limited. He used short, simple sentences consisting of a few words. He could read and repeat simple sentences and name common objects. When asked to count to twenty, he counted to eleven and stopped. He could not count backward from ten or say the alphabet. Henry could not spontaneously recall my name, but when I said, "My name is Suzanne; do you know my last name?" he replied, "Corkin." When I asked, "What do I do?" he responded, "Doctress." It was heartwarming to see that Henry, even in the throes of dementia and other maladies, still had his sense of humor; when asked, "You're not working anymore?" he replied, "No, that's one thing I'm sure of."

The positive news in 2007 was that Henry's epilepsy was stable; Bickford staff members did not observe any grand mal seizures. He continued to be friendly, pleasant, and conversational, and was a passive participant in group activities but fell asleep if not stimulated. He often listened to music in the lounge and watched TV in his room.

The last time I saw Henry alive was on September 16, 2008, during my annual pilgrimage to Bickford. As before, a neurologist accompanied me to document Henry's status. Prior to our visit, his Bickford doctor noted that Henry had declined significantly in the past year and that he was having a great deal of seizure activity. Henry was now eighty-two years old and still confined to his bed or geriatric chair. He was unable to feed himself and had difficulty chewing and swallowing food. Most of his communication was with gestures rather than words. When we saw him, he was sleepy but could be roused. He was essentially mute but attempted to say one or two words during our visit. Bickford staff members were very attached to Henry and were saddened to see his precipitous decline. I shared their sorrow.

It had been forty-six years since Henry and I first met at the Neuro. In the course of those years, he had been a regular presence in my life. We had watched one another age over the decades, although he did not know it. I had become used to his smile and kind manner, and I had heard his catch phrases and stories so many times that I could recount them myself verbatim. Many in my lab had been similarly touched by the experience of knowing Henry. He had permeated our culture, and we often found ourselves talking in Henryisms. For instance, if I asked a colleague whether she planned to attend a particular seminar that day, she might reply, "Well, there I'm having an argument with myself— should I go, or should I stay in the lab?"

One great gift that memory bestows on us is the ability to know one another well. It is through shared experiences and conversations that we form our deepest relationships—and without the ability to remember, we cannot watch these relationships grow. Although Henry had acquired many friends during his life, he was unable to feel the true depth of these connections. He could not get to know others well, and, tragically, he

285

could not know the lasting impression he had made on all of us who knew him—and on the world.

On my last visit, I stood beside him and said, "Hi, Henry. It's Suzanne, your old friend from East Hartford High School." He looked in my direction and gave me a faint smile. I smiled back. He died two and a half months later.

Thirteen

Henry's Legacy

Just before five-thirty on the afternoon of December 2, 2008, I received a call from the Director of Nursing at Bickford. Henry had died a few minutes earlier. I had just arrived home and was still sitting in my car when I got the news. Henry—the amiable, smiling man who had been a part of my life for so many years—was now gone. But at that moment, I had no time to mourn; Henry was dead, but he remained a precious research participant. It was time to put into action the brain donation plan that we had been organizing for the last seven years. My colleagues and I would have this one opportunity to study and preserve the most famous brain in the world. Our mission would be a challenging adventure, one that would allow no room for error. We would begin by scanning and harvesting Henry's brain. I knew I had a long, intense night ahead of me.

Throughout his life, Henry served science through his willingness to undergo countless tests and examinations. The postmortem research on his brain would be a beautiful finale to his enduring contributions. Henry gave us the rare opportunity to examine in death a patient whom we had studied extensively in life. MRIs are tremendously useful but imperfect; the only way to truly understand the nature of Henry's amnesia would be to look directly at his brain and document the damage. With MRI scans, we could estimate the lesion but could never characterize it with certainty; now, we would finally understand the anatomical under-pinnings of his amnesia.[1]

Before Henry, only a limited number of "classic" brains—that is, brains of patients whose cases represent historic insights about localization of function—had been studied postmortem. These previous cases yielded useful but limited information. Studying Henry's brain would give us the opportunity to make a pioneering contribution to the science of memory. Detailed planning by our dedicated team of researchers would allow us to unite five decades of well-documented behavioral research with the best possible technologies for brain imaging, preservation, and analysis, providing the most complete information yet on a single person's brain.

After receiving the news of Henry's death, while still sitting in my car, I called Jacopo Annese, a young researcher at the University of California, San Diego. He would be responsible for taking Henry's brain to San Diego for preservation and further study, in close collaboration with the Mass General team and me. We had agreed that when Henry died, Jacopo would travel to Boston to be present at the autopsy. As soon as I told him what had happened, he arranged to take the redeye to Boston.

I grabbed my purse, hurried up the stairs to my condo, and set to work. My colleagues and I had drawn up a flowchart of who needed to be contacted and in what order when Henry died. My assistant had laminated a wallet-size version of the flowchart for each of us to carry with us, and I kept copies under the wall-mounted phone in my kitchen, in my car, in my office, and on the desktop of each of my three computers. I grabbed the copy from my kitchen along with the laminated consent statement for Henry's autopsy, and set up shop on my dining-room table.

I first called the man who would remove Henry's brain—Matthew Frosch. At the time, he was interviewing a Harvard Medical School applicant so his phone was turned off. I paged him and he called back as soon as the interview was finished, guessing what had happened. He agreed to line up everything needed to perform the autopsy the following morning. I reassured him that I would obtain the legal permission for the brain donation.

Even though Henry and his conservator, Mr. M., had signed an Authorization for Brain Autopsy form, in 1992, I wanted to get Mr. M.'s consent after Henry's death for the autopsy and brain donation. In search of someone to witness the consent process, I rushed next door to my neighbor's house and rang the bell several times; she finally appeared. "I need a witness!" I blurted, and explained briefly what was happening. Without hesitating, she accompanied me to my house and stood next to me at the dining-room table, leaning in toward the phone. With a heavy heart, I called Mr. M. and told him that Henry had died that afternoon. I reached him on his wife's cell phone, interrupting their dinner out with their teenage granddaughter. As my neighbor listened attentively, I read Mr. M. the consent form, phrase by phrase, and he repeated them back to me. He gave permission for an autopsy without restriction, and for Mass General to use "all tissues and organs that have been removed" for research or to dispose of them in accordance with its policies. I thanked Mr. M. and told him of our plans to scan Henry's brain overnight before taking his body to the morgue in the morning.

Next, I called my assistant, Bettiann McKay, who was shocked when I told her Henry had died. She had known he was ill, but had not felt any sense that the end of his life was imminent; he had always bounced back after previous scares. I asked whether she would be willing to spend the night at my house and look after my three high-maintenance pets that needed care while I was overseeing the process of scanning and harvesting Henry's brain. When Bettiann arrived, she quickly took over the responsibility for the dogs and cat with her usual down-to-earth demeanor, but admitted later that she felt an onrushing sense of grief. She had recently spoken with Henry's nurses about what he would like for Christmas, and had ordered, wrapped, and sent a children's art set, picturing his surprised smile when he opened the gift. Bettiann and other lab members were planning to go down to visit with him before Christmas and take a small tree for his room. Even though she did not know

Henry well, and he never remembered her, she felt that he was part of the close-knit family our lab had become.

As the calls continued through our phone chain, our team planned on gathering at the Mass General Martinos Center, located in Building 149, which formerly served as a naval supply center. I was fortunate to have this internationally renowned imaging center two blocks from my home, and to be on the faculty there. This facility houses nine powerful MRI scanners as well as a series of other technologies used to image brain structure and activity. The center is also a leader in developing new methods to glean information from living brains.

At around quarter to six, André van der Kouwe, a biomedical engineer who directs the programming of the Martinos Center scanners, was informed of Henry's death. Soon after, André saw Allison Stevens, a young imaging researcher and part of our team, putting on her coat to leave.

"Aren't you staying for all the excitement?" he asked. She gave him a baffled look. For some reason, word had not yet reached Allison.

"H.M. died," he told her.

"What?" she yelled. "But what about the phone chain?"

Panicking, she began tracking down the rest of the team. She stopped to send a simple text message to the imaging researcher who had scanned Henry's brain in the past: "H.M. died." She also called one of my trusted graduate students, who told her that Henry's body was on the way, expected to arrive at about eight-thirty, and that we needed to get hold of a tarp to cover the scanner bed in case any bodily fluids leaked out. None of us had scanned a dead body before, and we wanted to be prepared for all eventualities.

By the time I arrived at the Martinos Center around eight, the imaging team had just finished eating dinner and was ready to spend the entire night scanning Henry's brain. His body, still en route from Connecticut, would arrive soon. I had asked the driver to call me when he neared the building, and when I realized I had no cell phone reception in the building, I went outside to wait for the hearse, huddled against the frigid

Boston weather in a full-length down coat, hood, and mittens. At around eight-thirty, I saw a vehicle rounding the corner through the darkness, moving hesitantly. I ran toward it, waving my arms over my head.

"I'm Suzanne Corkin! I think you're looking for me."

I guided the driver to the ramp of Building 149, where a Mass General police officer waited. My colleagues hurried out of the building to assist the driver in rolling the gurney inside. As Henry's body emerged from the hearse, I noticed that he had been covered with a patchwork quilt with a hood that covered his head, and another hood over his feet. Somehow, I felt comforted by this homey touch.

Luckily, the building was deserted, so no one would be alarmed at the sight of a dead body being wheeled across the atrium. Mary Foley, the technologist who keeps the Martinos Center running, had already gotten permission from security to transport Henry's body. She and Larry White were part of the team who had scanned Henry years before, and they were waiting in a suite of the building called Bay Four, which houses a scanner with a powerful three Tesla magnet. This suite would be the site of our all-night marathon to obtain many different kinds of MRI scans of Henry's brain.

The cylindrical scanner had a tunnel with a bed for the participant. Outside the scanner room was an anteroom, where technologists and researchers could view the scanner through a window while controlling it from a computer console. Before Henry entered the scanning room, we had to transfer his body to a nonmagnetic gurney; the scanner's powerful magnet could easily suck even large metal objects into its cavern. Concerned about Henry's large size, we had made sure that half a dozen strong men were available to lift him onto the scanner bed.

Beneath the quilt, Henry's body was encased in a black body bag over another clear bag. Our team unzipped the bags and peeled them down to expose his head and torso. We then removed the ice packs that had been placed around his head at Bickford to help preserve his brain tissue. For some of my colleagues in the room, this meeting with the famous H.M. was their first. For David Salat, who had known Henry in life, the reality of this moment hit him with his first glimpse of Henry's impassive face. Outwardly, we were calm and composed; inwardly, we were nervous, knowing this was a historic event in the annals of neuroscience, and we had one chance to get it right.

With a feeling of fondness, our team gently hoisted Henry's body to the stretcher. His illness had resulted in weight loss, so lifting him was easier than we had anticipated. We rolled him into the scanner room and transferred him to the bed. Mary pushed a button, and the bed glided into the bore of the magnet.

Even though we had scanned Henry's brain many times during his life, it was important to collect MRI images after his death—first *in situ* (with the brain still in the head) and later *ex vivo* (with the autopsied brain in a custom-made chamber). Scanning after death has several advantages. Living people who get MRI scans are instructed to lie perfectly still because any motion interferes with the quality of the scan, but even with a cooperative participant, MRI researchers must correct for natural motion—breathing, pulses of blood, and other minor movements. Now that Henry was dead, we would have no motion interference and would be able to obtain extremely clear images. Scanning living people is also limited by their tolerance for the procedure. They are confined inside the bore of the magnet, which often makes people feel claustrophobic and antsy, and even the most easygoing participants can tolerate the scanner for only two hours at the most. That night, without these impediments, we had the opportunity to scan Henry for nine hours, and we could gather an unprecedented amount of pristine data.

The ultimate goal of biomedical imaging is to give doctors and researchers detailed pictures of the particular body structures that are the focus of their treatments and research. The human body is not a map with clearly marked borders and road signs; it is often difficult to distinguish one tissue or cell type from another. MRI, however, enables us to do just that. When a participant enters an MRI scanner, he is exposed to a strong magnetic field, which causes the spins of the hydrogen nuclei in his body to line up with the magnet. The technician introduces radio frequency pulses into the field, briefly knocking the spins out of alignment with the magnet. As the spins realign with the magnetic field, they transmit a signal that is detected with a coil and used to create an image of the body. Additional magnetic-field pulses further manipulate the spins to encode spatial information in the images (to detect where tissues are located) and to vary the contrast in the images. The series of radio frequency and magnetic field pulses is called an MR sequence. MR sequences have distinctive sounds that the participant experiences in the magnet, and these sequences generate images with characteristic tissue contrast. Brain imaging is typically done with a variety of MR sequences designed to illuminate different properties of brain tissue—gray matter, white matter, cerebrospinal fluid, and the borders between brain structures.

One question we had was whether the sequence of radiofrequency pulses and magnetic gradients we would use to generate images of Henry's brain should differ from those typically used on a living body. A month earlier, I had given a short presentation describing Henry's case at the annual breakfast meeting of the Dana Foundation, which had given partial financial support for these planned postmortem studies. At the end of my talk, I asked anyone who had experience doing MRIs of dead brains to let me know what sequences they used. A colleague suggested that I contact Susan Resnick, who had been scanning the brains of corpses as part of the Baltimore Longitudinal Study on Aging. Before we started scanning, I phoned Susan and was relieved when she told me she used the same sequences for dead brains as for living ones. At least we could obtain useful data from routine clinical scans before proceeding with more experimental studies.

Our MRI team began with standard scans that would be performed on a living patient in a clinic, and then progressed to increasingly high-resolution scans that would show the anatomical details of Henry's brain, first at the level of a millimeter and finally down to a few hundred micrometers, showing large groups of brain cells. As the images began to emerge, André was struck by their beauty: the borders of brain structures were unusually sharp. At the highest resolution, even the walls of tiny blood vessels in the brain—usually distorted by the movement of blood—were easily visible in the stillness of death. He could also see clearly the gaping holes on the left and right sides of Henry's brain, where the lesions were.

As my colleagues gathered data from the MRI study, I turned my attention to other pressing matters. A nearby funeral home in Charlestown had agreed to transport Henry's body the short distance from Building 149 to the Mass General morgue in Boston, where his brain would be removed. The funeral home balked at the last minute, however, saying it would not move a body without a signed death certificate—the nursing home had rushed his body to us before getting a signature. We worried that if the autopsy were delayed, the brain would lose its firmness and become difficult to extract and preserve. At one point, we half-joked about whether we would have to resort to putting Henry on a stretcher and rolling him ourselves across the bridge to the hospital's main campus. Then, in the same spirit, I remembered walking past a funeral home in the North End. I called the home and, although it was now late at night, a man answered. In my most professional voice, I explained that I needed to transport a body from the Charlestown Navy Yard to the Mass General morgue. He said he would consult his boss and get back to me. A few minutes later, the boss called back and agreed to have a hearse there by early morning.

By dawn, my colleagues had generated eleven gigabytes of brain images. To put this amount of information in perspective, a typical MRI scan with a living participant generates a few hundred megabytes of data, enough to fit comfortably on one CD. We would need sixteen CDs to hold all the information we had collected from Henry's brain that night. We were lucky that the scanner—a finicky machine with frequent mechanical problems—had held up throughout the nine-hour session. As it turned out, it broke down mere hours later.

Henry's body needed to be out of the building by six a.m., before hundreds of researchers began filing in. The hearse arrived at five-thirty; at six, with Henry's body strapped on, we rolled the gurney through a backdoor and down a ramp to the hearse. With the hearse en route to the Mass General morgue, I jumped in my car, accompanied by a former lab member, then a Tufts University medical student. We hustled to Logan Airport to pick up Jacopo, who by then had arrived from the West Coast. On his flight, Jacopo later told me, he reread some of the seminal papers on Henry and mentally rehearsed the autopsy procedure he had been taught as a graduate student. He outlined the details of his plan to create, preserve, and disseminate an anatomical library of large-format glass slides and digital images, together representing Henry's entire brain.

On the return trip from Logan, we stopped at Starbucks to fortify ourselves with cups of espresso, and then headed to Frosch's Mass General office. The autopsy would be an indelible learning experience for all of us. Henry's body was safely stored in a large refrigerator in the basement of the hospital's Warren Building. We gave Matthew a CD with some of the beautifully clear images André had obtained the night before. These pictures gave Matthew guideposts for the removal process. He was concerned about the site of Henry's surgery: scarring might have caused the brain to stick to its cover, the dura, the tough layer between the brain and the skull. It would be difficult to remove the brain without leaving fragments behind. Losing tissue from the surgery site would hinder our ability to provide definitive evidence of the extent of tissue missing from Henry's brain. As Matthew looked through the scans, he was reassured to see that a good amount of fluid remained between the brain and the dura in that region. He printed several images to use as a reference in the autopsy room.

Down in the Warren basement, the senior pathology technician wheeled Henry into the autopsy room. Matthew, Jacopo, and the medical student joined him, along with a photographer from the Pathology Department who would document the procedure. So as not to get in the way, I stayed in the "clean room" next door, behind a huge glass window through which I could watch the delicate procedure unfold. I stood on a chair, eager to get the best possible view.

Matthew first made a shallow cut across the top of Henry's head, from behind one ear to behind the other. He then peeled back the scalp in both directions to expose the skull. At the front of the skull, they could see the faint outlines of the two burr holes that Scoville had drilled decades ago, just above the ridges of Henry's eyebrows. These plugged holes had healed well, so Matthew was able to cut around them. The next task was to remove the top of the skull, which is the trickiest part because the dura tends to stick to the surface of the skull, especially in older people. The technician made the first pass, using an electric saw to cut all the way around the head at one level, going only partway through the bone. Matthew, a highly experienced neuropathologist, deftly completed the cut without nicking the brain. He then used a chisel to pry off the skull. To our relief, it lifted off smoothly. Although Matthew seemed to proceed with complete confidence, he later admitted that he had deliberately turned his back to the window so that I would not see how much he was sweating.

Matthew pulled back the dura, starting at the frontal lobes. He next lifted the frontal lobes to loosen them from the skull, severed the optic nerves to disconnect the brain from the eyes, and cut the carotid arteries to detach the brain from the circulatory system. The brain was now freed enough from its moorings that he could move it from side to side to view the site of the surgery on either side. He noticed areas where the dura was stuck to the brain, particularly on the right side. Taking a fresh scalpel, he carefully cut the dura from these sections. He then worked to free the back of the brain, a job that was facilitated by the shrinkage of Henry's cerebellum, caused by Dilantin. Matthew was now able to lift the intact brain out of the skull and place it in a large metal bowl.

At one point during the autopsy, I had left to call Brenda Milner who, at age ninety, was still working at McGill. I told her Henry had died; it was not unexpected news, and she accepted it with equanimity. I asked her not to tell anyone yet, as I wanted the autopsy to be completed before we announced Henry's death to the world and started fielding calls and emails from the press and the scientific community. When the brain was out and intact, I called Milner again to report our success. Seeing Henry's precious brain in the safety of the metal bowl was one of the most memorable and satisfying moments of my life. Our planning for this day had evolved over years, and we pulled it off without a hitch. Those of us who witnessed it were elated and smiling; I raised my hands over my head and applauded Matthew.

Matthew transferred the brain into the clean room by passing it through a refrigerator that had a door in each room. The entire neuropathology staff joined us in the clean room, and we all took a good look at the brain while the photographer snapped shots from all angles. Matthew then tied a thread around the basilar artery and secured the thread to the handle of a bucket full of formalin, thus allowing the brain to float in the solution without sinking to the bottom and becoming distorted. The solution would change the brain from its soft tofu-like consistency to something more firm like clay. A few hours later, Matthew transferred the brain to a special formaldehyde solution. He had bought the concentrated paraformaldehyde in advance, stored it in his cold room, and made a fresh bucket's worth of fixative that morning.

After taking Matthew and Jacopo to lunch, it was time to let the world know of Henry's death. During scanning the night before, I sat in the control room and typed out his obituary on my laptop. It was the first opportunity I had had to stop and reflect on Henry's death, and it was also a chance to sum up briefly his enormous contributions to the science of memory. I drove to my office at MIT and emailed the obituary to faculty members in my department, to former lab members who had worked with Henry, and to Larry Altman, a veteran medical reporter at the *New York Times*. He passed it along to reporter Benedict J. Carey, who wrote an elegant obituary that appeared on the front page of the newspaper two days later, December 5, 2008. The article brought the attention of the wider public to this case so well known in the world of neuroscience. For the first time, Henry's name was made public, and with the conservator's permission, we announced to the world that "the amnesic patient, H.M.," was Henry Gustave Molaison.[2]

Ten weeks later, in February 2009, Jacopo returned to Boston, bringing a custom-made Plexiglas chamber that would hold the preserved brain during a new round of imaging outside the body, *ex vivo* imaging. The

Martinos imaging team first used the same three Tesla scanners they had used the night Henry died. This set of scans would provide a bridge between the geometry of the brain as it was in life with the final form it would take after being cut into ultrathin slices. The brain tissue could be further stretched or reshaped when sliced and placed onto glass slides. The MRI imaging data would allow us to measure and correct any deformities in the brain slides so they could be mapped back to the original architecture of Henry's brain.

In addition, we had the opportunity to image the brain in a scanner with a seven Tesla magnet—one of the strongest magnets currently used with humans. It would provide a more accurate and detailed picture of the human brain than most researchers have ever seen. We did not put Henry in this scanner the night he died because his brain retained two metal clips used to tie off blood vessels in his operation, and we feared that they would heat up in the powerful scanner, creating further damage to his brain. Matthew removed the clips at the autopsy, so they were no longer a concern in the preserved brain. Jacopo had worried that the heat generated by the seven Tesla magnet could harm Henry's brain, even in the absence of the clips. To reassure him, Allison had run trials with other brain tissue and estimated that the tissue would not heat up more than about three degrees, which was perfectly safe.

The seven Tesla magnet had a smaller head coil (the opening for the head) than the three Tesla. A technical challenge, therefore, was to craft a chamber that was small enough to fit in the head coil but large enough to hold Henry's brain wrapped in formaldehyde-soaked cotton for protection. The chamber that Jacopo brought from San Diego was the right size, but the cotton surrounding Henry's brain collected air bubbles. The trapped bubbles turned out to be a technical problem in the *ex vivo* images because the bubbles appeared larger than their real physical size and obscured the adjacent brain tissue. Unfortunately, these artificial blobs occurred in some interesting temporal-lobe regions.

After three separate scanning sessions over a long weekend, the imaging team had finally gleaned all it could from Henry's brain, and it was time for it to travel to the University of California, San Diego, for cutting. On February 16, I met Jacopo at the Martinos Center, where he was waiting in the vestibule guarding a cooler that contained Henry's brain in its ice-encased chamber. A PBS film crew joined us to document our journey from Mass General to the door of the plane. We all climbed in a van with the producer in the front passenger seat, her assistant driving, and the cameraman in the middle seat facing backward so he could film Jacopo and me in the backseat—the cooler tucked snugly between us for protection.

When the van pulled up to the curb at Logan Airport, a welcoming committee was on hand—representatives from the Transportation Security Administration and JetBlue Airways, and the Director of Communications at Logan. I knew it was critical to pave the way for this unusual piece of carry-on luggage, so a month earlier, in a letter to the Customer Support and Quality Improvement Manager in the US Department of Homeland Security at Logan, I had requested help in transporting a human brain from Boston to San Diego. I explained how the brain would be packaged, that Jacopo would accompany it on the flight, and that upon landing he would take it to his laboratory at the university. The chair of Jacopo's department also wrote a letter, confirming that Jacopo was a faculty member in his department and underscoring the extreme importance of the mission. Walking through the airport, we felt like celebrities: the film crew tracked our path, and people stared, wondering who we were and why we were being filmed. At the security checkpoint, a uniformed woman approached us and gave us the welcome news that she would carry the cooler around to the other side, so as not to expose the brain to radiation. Relieved, we passed through security in the normal manner and retrieved the cooler.

When it was time to board, Jacopo and I enacted a formal exchange for the camera. I carried the cooler to the door of the gate and placed it on the ground. We smiled at each other and hugged; he then picked up the cooler and walked down the ramp, stopping once to turn around and wave. At face value, Jacopo was just a scientist carrying a cooler with a brain that had been fixed in formaldehyde. But what he was carrying remained precious to me. I felt sad to see Henry's brain go—it was my last goodbye to him.

As I left the gate with the PBS crew, I looked out at the plane. Henry's most memorable experience from his childhood was that half-hour plane ride over Hartford. If only he knew that his last trip would be a twenty-five-hundred-mile flight in a big jet, he would have been ecstatic. The moment had an air of finality about it.

On December 2, 2009, a year after Henry's death, I stood in a laboratory at the University of California, San Diego, where Jacopo was preparing to cut Henry's brain into slices as thin as a strand of human hair—seventy microns. Normally, brains used in research are cut into large sections or slabs and then sliced thin enough for the tissue to be viewed under a microscope. Henry's brain would be cut whole, from front to back, with the goal of collecting full vertical planes through the brain, rather than isolated pieces. The brain had been immersed in a solution of formaldehyde and sugar. The sugar soaked into the brain tissue and to prevent ice crystals from forming when the brain was frozen in preparation for the cutting. Before freezing, it was placed in a mold filled with gelatin, which would help this precious organ hold its shape. It was important to maintain a delicate temperature balance throughout the cutting—keeping the brain cold enough that the blade could slice the tissue cleanly, but not so cold that the tissue would shatter.

Everyone in the lab was excited and anxious as the process began. During the cutting, which lasted fifty-three hours, visitors appeared from time to time. Benedict Carey from the *New York Times* flew out to capture this seminal moment in memory research. Jacopo had also invited several luminaries at his university to view the event, including the famed neurologist Vilayanur S. Ramachandran, neurophilosophers Patricia and Paul Churchland, and eminent neuroscientist Larry Squire. Because the cutting went on endlessly, the conference room was decked out with treats for the lab members and visitors to enjoy—platters of food and a delicious Italian cake. Jacopo had hired a film crew to document the entire procedure. They positioned cameras in the cutting room to stream the events live on the Web, and over the course of three days, 400,000 people visited the site to witness this historic undertaking.

For the cutting, the brain, embedded in its block of frozen gelatin, was secured on an electronic device, a microtome, which acted like an exceedingly precise meat slicer. To keep the brain cool, technicians pumped liquid ethanol through a tube into the surrounding spaces. Jacopo, wearing black gloves, sat in front of the microtome. Each pass of the blade over the block of ice revealed a delicate roll of brain tissue and gelatin, which he gently wiped up with a large, stiff paintbrush and placed in a well in a partitioned container that resembled an ice-cube tray, filled with solution. The front of Henry's brain faced upward, and a sixteen-megapixel camera mounted above captured and numbered each surface before it was cut. Each slice was placed in a well with the corresponding number. The cutting began at the front of the brain and proceeded back—from the frontal pole to the occipital pole. As exciting as the project itself was, the painstaking work of cutting and securing thousands of brain slices was monotonous. Still, the lack of drama meant that everything was going according to plan.

As of December 2012, Henry's intact brain has been examined by the neuropathologist at Mass General, scanned by investigators in the Mass. General Martinos Center, and cut into seventy-micron slices at the University of California, San Diego. My colleagues and I continue to advance the coordination of these different studies—both to allow for a final diagnostic neuropathological examination at Mass General and to address the numerous research questions that are waiting to be answered.

Once this work is underway, we will learn with certainty which medial temporal-lobe structures were preserved in Henry's brain and to what extent. Although the residual nubbins of hippocampus and amygdala were nonfunctional, the remaining portions of the neighboring cortex—perirhinal and parahippocampal—may have been working. Knowing the status of this residual memory tissue will help to explain Henry's unexpected knowledge, such as his ability to draw the floor plan of the house he moved to after his operation. We are also eager to know the effect of his lesion on the structure and organization of areas that were connected to the medial temporal lobes—the fornix, mammillary bodies, and lateral temporal neocortex. Some brain areas beyond medial temporal-lobe structures are known to support declarative memory in normal individuals, so a further question concerns the structure and organization of those areas—the thalamus, basal forebrain, prefrontal cortex, and retrosplenial cortex—and of preserved nondeclarative memory areas—the primary motor cortex, striatum, and cerebellum. Henry's MRI scans told us that his cerebellum was severely atrophied, and now we can document the specific areas that were affected.

The 2,401 slices of Henry's brain have been frozen and are currently stored in a protective solution. Some will be placed on large glass slides, about six inches by six inches, and stained using various methods to reveal details about the cells or the anatomical boundaries of brain structures, such as those around his lesion. Some of these sections will be stained with dyes developed by neuropathologists in the nineteenth and early twentieth centuries to show normal brain structure—identifying the neurons, highlighting their organization and connections, and revealing the white-matter tracts that connect one brain region to another. Other sections will be stained using methods from the late twentieth and early twenty-first centuries that engage antibodies—proteins that detect the abnormal proteins that mark diseases such as Alzheimer and Parkinson. With this combination of approaches, a careful analysis of the abnormalities in Henry's brain tissue will reveal a vast range of new information. Future research will tell us the kind of dementia he had when he died, the exact locations of his small strokes, and the consequences of his surgery—both for the regions adjacent to the operation site and for distant areas that had been connected to the removed structures.

The digital images captured during the cutting will be used to create a massive three-dimensional model of Henry's brain, which will eventually be available on the Internet for anyone to explore. This brain will be the centerpiece in the Digital Brain Library Project at UCSD, which aims to collect and archive the brains and personal profiles of one thousand individuals over the next decade. Even after death, Henry will continue to make groundbreaking contributions to science.

Henry's legacy has many layers. From studying him directly, we have amassed the largest and most detailed collection of information on a single neurological case—Scoville and Milner's legendary research, reams of test results collected over decades, our descriptions of his everyday life, the imaging of his intact brain in life and death, and the precious slices of his physical brain. This astonishingly large dataset representing a single brain would in and of itself secure Henry's legacy as a vital contributor to the history of neuroscience, but his impact goes beyond that. His case inspired thousands of researchers to investigate other kinds of amnesia and disorders related to memory loss. In addition, what we learned about Henry motivated generations of basic scientists to study memory mechanisms, creating a myriad of different approaches with non-human primates and other animals. These huge advances allowed researchers to explore all manner of issues in basic and clinical science. Henry's case triggered an extraordinarily fertile period in memory research, and the momentum continues to grow.

Epilogue

A week after Henry died, Mr. M. and his wife arranged for a funeral at Saint Mary's Church in Windsor Locks, Connecticut, not far from Bickford. Henry's body had been cremated. At the front of the church, an urn holding his ashes rested on a white pedestal surrounded by flowers. The front of the urn was etched with a cross and this message: "In loving memory, Henry G. Molaison, February 26, 1926—December 2, 2008." Nearby, a framed collage of photos gave a condensed glimpse of his life: Henry as a little boy, seated on a chair with one leg tucked under him, smiling; a dignified sepia-toned portrait from his twenties; Henry as an older man with white hair, sitting in his wheelchair wearing a white shirt and tie; and images of Henry's family and his younger days.

The ceremony was small, with only those close to Henry in attendance. I was privileged to deliver Henry's eulogy, in which I recounted the story of his operation and the groundbreaking studies that followed. I also spoke about the people who made it possible for Henry to make his contributions to science, including Lillian Herrick; her son Mr. M., who had taken over the responsibility of seeing to Henry's welfare; and the staff at Bickford. "All of these good people lit up his life," I said. "And he in turn illuminated other people's lives." I talked about Henry's personal qualities—his clever sense of humor, his intelligence, and his signature catch phrases.

"For many of us, losing him was like losing a family member," I concluded. "My colleagues and I are honored to have been part of his inner circle. Today, we say goodbye to him with respect and with gratitude for the way in which he changed the world and us. His tragedy became a gift to humanity. Ironically, he will never be forgotten."

After the funeral mass, we all made our way next door to the church hall for a reception; a staff member from Bickford and my assistant brought sandwiches and cookies. Others at the service included several Mass General colleagues who had recently taken part in the effort to image and preserve Henry's brain. For them, it was a chance to pay their respects, drink a little tea or coffee, and decompress after a frantic and sleepless week. Three former lab members who had worked with Henry were there, as were Bickford staff members and fellow patients.

After the reception, we drove to a cemetery in East Hartford for the burial, and walked across the flat expanse of grass to a large headstone at the site where Henry's parents were buried. Below their names were his name and date of birth; we could now fill in the date of death. The funeral director had prepared the gravesite, and the urn containing Henry's ashes was placed on a short, white Grecian column. We stood in a small semicircle around the urn while the church deacon said the words of committal. When he asked that we join him in prayer, we bowed our heads and thought about Henry.

The day after Henry died, I sent out a brief email sharing the news with other memory researchers. They forwarded the message to others, so the word spread quickly to scientists throughout the United States and Europe. Over the next few weeks, I received heartwarming thoughts from colleagues around the world, offering condolences and messages of praise for Henry. I also entertained requests from the media for interviews and articles about Henry.

Some colleagues replied with statements about Henry's contributions to science. A professor in the Psychology Department at Yale University wrote to Brenda Milner and me, "Learning of your work with H.M. was one of the major influences on the direction of my own thinking about cognition and memory very early in my career." Several faculty members at other universities mentioned that they would include a tribute to Henry in their classes that day, and former lab members shared anecdotes about him. I learned, for instance, that Sarah Steinvorth had sat with him in his room at Bickford and watched a John Wayne movie through to the end, and that Henry had been "on fire" the whole time, repeating, "I know this, I know this," and talking about his own gun collection. She said he was excited for some time after the movie ended.

I also learned that a former technical assistant had once played a practical joke on one of my graduate students by taking advantage of Henry's ability to keep items in his memory for longer periods of time by rehearsing them. She wrote:

> You recall, Henry was always game for a good joke. I told him that the next person coming in to test him was named John, and asked him to act surprised when he entered the room, and say "Oh, hi John," like he recognized him. We practiced for a few minutes, and then I ran like mad to bring John in for his session with Henry. Henry did an extremely believable double take, and as natural as could be, delivered his line perfectly. The look on John's face was priceless! Henry and I had a good laugh about it.

I had a forty-six year bond with Henry. While I did not speak about him in a sentimental way, he had grown to mean something to me. An MIT historian captured my sentiments in an email to me after Henry's death: "This has to be a grievous loss for you. It is such an unusual relationship that it is hard to articulate its meaning—but surely it is true that you made an enormous difference in his life—as he did in yours." My interest in Henry, however, had always been primarily intellectual; how else would I explain why I had stood on a chair in the basement of Mass General, ecstatic to see his brain removed expertly from his skull? My role as a scientist had always been perfectly clear to me. Still, I felt compassion for Henry and respected him and his outlook on life. He was more than a research participant. He was a collaborator—a prized partner in our larger quest to understand memory.

Over the years, as Henry lost his father and mother, and as he aged and his health became more vulnerable, my colleagues and I became the people who knew him and cared about him. The sense of family that we cultivated in the lab extended to Henry. We sent him cards and gifts, celebrated his birthdays, and had his favorite foods on hand when he visited. I looked after his medical care and found a dependable and caring conservator for him. Although Henry could not remember, I take comfort in knowing that in the days he spent working with my colleagues and me, he knew that we were learning from him and that he was special. That knowledge was gratifying for him and gave him a sense of pride.

Henry's legacy extends beyond science into the realm of art and theater. In 2009, shortly after his death, Los Angeles artist and filmmaker Kerry Tribe created a 16-mm film installation, *H.M.* The film explored Henry's case using actors, interviews with me, images of the apparatuses used in our experiments, and photographs of Henry's world. During the screening, a single reel of film passed through two adjacent projectors, showing the identical picture on each of two screens with a twenty-second delay, mimicking the duration of Henry's short-term memory. Tribe's innovative film was featured in the 2010 Biennial at the Whitney Museum in New York, and Holland Carter of the *New York Times* called it "extraordinary." The same year, Marie-Laure Théodule created a seven-page graphic story about H.M., which she published in the summer edition of a French scientific publication, *La Recherche*. She accurately described his operation and the subsequent research but transformed Henry into a thin, dapper-looking gentleman in a suit, dress shirt, tie, and hat. In 2010, New York–based psychologist and playwright Vanda premiered *Patient HM*, which dramatized her intuition about what Henry was like as a human being. The next summer, the 2011 Edinburgh Festival featured a play entitled *2401 Objects*, referring to the 70-micron slices of Henry's brain. This production by the Analogue theater company told the touching story of Henry's preoperative and postoperative life. *Scientific American Mind* featured a one-page graphic story in July 2012, which accurately conveyed the scientific messages in Henry's case.

The Internet is home to a growing community that is fascinated by Henry and his story. When we Google *Henry Molaison*, more than sixty-two thousand results appear. Henry Molaison is the topic of a Wikizine (an interactive magazine created and edited by users). A blog called *Kurzweil Accelerating Intelligence* has a discussion of H.M.'s case, and other sites, such as *Amusing Planet* and *Brain On Holiday,* devote pages to Henry. The astonishing widespread interest in Henry is a fitting testimony to his unforgettable life.

My work with Henry often focused on the details of measuring behavior and interpreting data, but of course his case raises larger questions for society. How are we to see the life of Henry Gustave Molaison? Was he purely a tragic victim who lost an important part of his humanity to medical experimentation, or was he a hero for furthering our understanding of the brain?

The more I consider Henry's case, the more difficult it seems to answer these questions. No neurosurgeon today would perform the same operation that Scoville did on Henry, and indeed Scoville himself warned others against attempting the same procedure after its results became clear. But unlike the more dubious practices of psychosurgery, such as prefrontal lobotomy and bilateral amygdalectomy, Henry's bilateral medial temporal-lobe resection was intended to alleviate a specific debilitating disease, and it did decrease the frequency of his seizures. Moreover, Henry's operation was nested in a long, fruitful medical tradition of experimental procedures.[1]

Doctors and patients often face tough choices, but neurosurgeons universally agree that a procedure that has a known devastating effect, such as wiping out a patient's memory, should not be performed. Wilder Penfield touched on this idea in a paper discussing the two cases of amnesia in his own patients, F.C. and P.B. "As a surgeon I take this responsibility very seriously," he said. "I know that Dr. Scoville does likewise. We must always balance the dangers of disability and of death against hope of helping our patients." In 1973, twenty years after Henry's operation, Scoville wrote about psychosurgery in the *Journal of Neurosurgery*: "If destructive surgery benefits overall function, it is justified; if overall function is made worse by operation, it is not justified."[2]

Was becoming amnesic an acceptable price to pay for seizure control? Most would agree that the answer is decidedly no. Yet it is not clear that Henry would have lived to the age of eighty-two had his seizures continued as they had prior to his surgery. The seizures themselves could have produced devastating outcomes: in the extreme, Henry could have died as the result of an injury sustained during a seizure. Moreover, drug-refractory epilepsy patients often have abnormalities of the heart and blood vessels, which sometimes result in sudden death, and evidence also suggests that repeated seizure activity causes damage to neurons. Further, Henry's seizures could have compromised his breathing and other vital functions, possibly resulting in death. Or he might have been one of the unfortunate patients who goes into *status epilepticus*—thirty minutes or more of continuous seizure activity. This life-threatening occurrence is considered a medical emergency, and despite aggressive treatments, patients sometimes die during status epilepticus due to heart failure or other medical complications. In this sense, Henry's surgery had significant benefits. So, although his quality of life was severely compromised by his amnesia, he probably lived a much longer life than he would have had his seizures continued at the preoperative frequency. Scoville arguably saved Henry's life, even if he took his memory.[3]

With the benefit of hindsight, no one would perform Henry's operation today, but was Scoville justified in performing it in the first place, when its outcome was unknown? In many cases, medicine advances through the willingness of patients and doctors to take risks. Those gambles may be relatively minor, as in agreeing to participate in a clinical trial of a drug that has been thoroughly tested for safety in animals and humans. Other times, the patient's decision requires a much more dramatic leap of faith. Operations we now consider routine—organ transplants, artificial heart implants, coronary bypasses—all depended initially on volunteers to take part in experimental procedures.

Risk is inherent in every kind of surgery, and risks intensify when complex and fragile organs such as the brain are concerned. Stricter medical ethics codes, our increasingly litigious society, and the rise of bioethics as a formal discipline have made the public and the medical community aware of the need to scrutinize the justification for daring procedures. We now have much better knowledge of individual brain structures and their roles, and a more realistic sense of what brain surgery can and cannot do in terms of alleviating psychiatric or neurological disorders. Still, experimental procedures continue to raise ethical questions. While rules for new treatments and devices are far more stringent now than at the time of Henry's operation, highly experimental surgeries are not as formally regulated, and surgeons sometimes make decisions for an individual patient, without the benefit of data from large clinical trials or animal studies.

Countless patients, such as Henry, have undergone procedures with the knowledge that the outcome was uncertain. Sometimes these people ended up benefiting society in ways that could not have been anticipated. Henry's chief motivation for participating in research after his operation was to help other people—and he did. For instance, after his death, I received a note from a woman who had temporal lobe epilepsy and had read about H.M. She had weighed the possibility of having her left hippocampus and amygdala removed to alleviate her seizures. Hundreds of patients with intractable epilepsy have benefited from surgery in which one temporal lobe was partially removed. Henry showed us that removing the hippocampus on both sides of the brain would cause irreparable loss of memory functions. The amnesic patients whose cases I described, F.C. and P.B., had a similar devastating outcome when their left hippocampus was removed surgically and their right hippocampus was already damaged. In order to prevent further tragedies of this sort, many candidates for epilepsy surgery now undergo a test that temporarily inactivates one side of their brain, allowing physicians to examine the integrity of language and memory separately on each side. This procedure—formerly called the Wada test and now called the eSAM (etomidate speech and memory) test—prevents surgical misfortunes. For example, if the patient makes errors on a memory test with the left side inactivated, the surgeon would not remove the right hippocampus because that would result in a bilateral hippocampal lesion.

"As I sit in my fourteenth floor office," the woman wrote to me in 2008, "with my window facing the Connecticut River and Hartford

Hospital, where Mr. Molaison had his surgery the year I was born, I am saddened by his death and grateful for the knowledge he provided. Because of him, my neurologists knew to do the Wada test to ensure that the hippocampus in my right temporal lobe was functioning prior to removing the other." She had a left temporal lobectomy in 1983 and remains seizure-free today.

Henry's case did not just help other patients; it also ignited the careers of countless neuroscientists. I interviewed a distinguished neurologist and geneticist at Children's Hospital Boston concerning the future trajectory of memory research. At the end of the interview, he was bursting to tell me how Henry influenced his life's work. "Like so many other neuroscientists, a big reason I'm in neuroscience is because of H.M.," he said. "I went to a small liberal arts college, Bucknell University, and I had the tremendous good fortune that Brenda Milner came and gave a seminar while I was an undergraduate. I was taking physiological psychology at the time. And so she actually came and lectured in our physio-psych class. It's never left me. That's a big reason why I'm interested in memory disorders and cognitive disorders—trying to understand how we can get at the mechanisms of memory."

Henry participated in a period of incredible change and advancement in our understanding of the brain, although he was unable to remember any of it. When he first became a subject of scientific inquiry, brain imaging was nearly nonexistent, and we collected data by hand with paper and pencil. During the 1980s, our cognitive testing became largely computerized, and in the 1990s MRI scans allowed us to visualize the structures and functions in his brain. By the time Henry died, we could study his brain with even greater precision. When our research with Henry began, we were all housed in psychology departments, and neuroscience was not even a full-fledged discipline. By November 2010, nueroscience had grown into a formidable discipline, and more than 30,000 neuroscience from all over the world attended the fortieth annual meeting of the Society for Neuroscience in San Diego. I had been invited to deliver a talk about Henry's contributions to the science of memory—a fitting way to celebrate his life.

The array of technologies now available to neuroscientists is staggering. Researchers can probe the molecular interactions within and between neurons, see the activity of large-scale networks in the living brain, scan the genome to find the genetic basis of neurological disease, and build complex computer models of brain structure and function. With all the tools we now have to study cells and gather large sets of data, we should remember how much can be learned by applying those tools to a single individual. By carefully examining one patient over time, we can fill in the gaps in our knowledge about how individual brains function and change throughout life, in health and disease. Our work with Henry provides an example.

Henry's case was revolutionary because it told the world that memory formation could be contained in a specific part of the brain. Before his operation, physicians and scientists acknowledged that the brain was the seat of conscious memory, but had no conclusive proof that declarative memory was localized to a circumscribed area. Henry provided us with *causal proof* that a discrete brain region deep in the temporal lobes is absolutely critical for converting short-term memories into enduring ones. Scoville's operation caused Henry to lose this capacity. Based on decades of research with Henry and numerous patients who volunteered their time and effort in our laboratory and many others around the world, we now know much more: short-term and long-term memory are separate processes that depend on different brain circuits; remembering unique events (episodic memory) and remembering facts (semantic memory) are both impaired in anterograde amnesia; learning with awareness (declarative memory) is impaired in amnesia, whereas learning without awareness (nondeclarative memory) typically is not. We also understand that a healthy hippocampus is essential for vividly recounting the details of a wedding (recollection), but that it is not essential for simply recognizing a face, without identifying it or placing it in a context (familiarity). Henry further showed us that the ability to recall and recognize information stored before the onset of amnesia differs depending on whether it is episodic or semantic information: most details of unique events are lost (episodic, autobiographical memory), but general knowledge of the world is preserved (semantic memory). Henry's case also underscored the value of donating one's brain for further study after death—a vital way for researchers to test their hypotheses and speculations, based on living patients, about the brain substrates that are responsible for specific learning and memory processes.

Amazing technological progress since 2005 has made it feasible to map the cognitive and neural mechanisms that underlie memory formation at the level of individual brain cells. The discipline of neuroscience is experiencing a series of transformative events driven by advanced technology. We can now observe with greater specificity the mysterious happenings inside the living brain. Sophisticated techniques will provide new kinds of information: optogenetic technology for precisely controlling specific neurons using genes, molecular engineering for fast and direct readout of neural activity, and connectomics to map the 100 trillion connections that make up the brain's neural networks. In parallel, cognitive scientists continue to advance theories about the fractionation and organization of memory processes, inviting researchers to map precisely defined computations onto discrete brain circuits.

Although each of these innovative technologies is fascinating in its own right, more important is what they can accomplish collectively. After spending decades mapping the overall anatomy of the brain and accumulating information at several levels, from behavioral to cellular, scientists are now striving to connect all that information into a comprehensive picture. In the field of memory research, we want to know how something so intangible as a thought or fact can lodge itself for decades in the living tissue of the brain. The ultimate goal of neuroscience is to understand how the billions of neurons in the brain, each with roughly 10,000 synapses, interact to create the workings of the mind.

We will, of course, never fully achieve that goal. Even as I type these words, I wonder what exactly is going on in my overcrowded brain. How do my networks of neurons marshal together the pieces of complex technical information I have learned, synthesize them into thoughts and perspectives, and put the total sum into words my fingers are then directed to type? How remarkable that the brain can fashion simple sentences out of such chaos. We will never have a formula to fully explain how the noisy activity of our brains gives rise to thoughts, emotions, and behavior. But the magnitude of the goal makes pursuing it all the more exciting. This challenge attracts brilliant adventurers and risk takers to our field. And even if we will never completely understand the way the brain works, whatever small part of the truth we are able to learn will bring us one step closer to understanding who we are.

Acknowledgments

Henry Gustave Molaison was the subject of wide-ranging experimental scrutiny for more than five decades. This research began in 1955 in Brenda Milner's laboratory at the Montreal Neurological Institute and moved to MIT in 1966. From 1966 until 2008, one hundred twenty-two physicians and scientists had the opportunity to study Henry, either as members of my lab or as our collaborators at other institutions. We all understand what a rare gift it was to work with him, and we are profoundly grateful for his dedication to research. He taught us a great deal about the cognitive and neural organization of memory. The research with Henry described in this book draws from these five decades of investigation.

During Henry's fifty visits to the MIT Clinical Research Center he received VIP treatment from many nurses and from the diet staff, headed by Rita Tsay; they deserve high praise for the wonderful care they gave him. For the last twenty-eight years of Henry's life, he lived at Bickford Health Care Center where he was affectionately looked after. Rich information about Henry's activities came from staff members at Bickford, and their accounts greatly enriched my telling of his story. Every time I had the slightest question about Henry, Eileen Shanahan provided an answer, and I thank her for that. Meredith Brown did a superb job of sorting through twenty-eight years of details in Henry's Bickford charts and summarizing the important points.

My gratitude for crucial suggestions and corrections goes to Paymon Ashourian, Jean Augustinack, Carol Barnes, Sam Cooke, Damon Corkin,

Leyla de Toledo-Morrell, Howard Eichenbaum, Guoping Feng, Matthew Frosch, Jackie Ganem, Isabel Gautier, Maggie Keane, Elizabeth Kensinger, Mark Mapstone, Bruce McNaughton, Chris Moore, Richard Morris, Peter Mortimer, Morris Moscovitch, Lynn Nadel, Ross Pastel, Russel Patterson, Brad Postle, Molly Potter, Nick Rosen, Peter Schiller, Reza Shadmehr, Brian Skotko, André van der Kouwe, Matt Wilson, and David Ziegler. Their sharp minds gave me brilliant and frank feedback, which greatly improved the book.

I benefitted enormously from stimulating discussions with colleagues in neuroscience who graciously agreed to let me record their views on the importance of Henry's contributions and the future direction of memory research. I intended to weave this exciting material into chapter 14, but alas it ended up on the cutting room floor. Nevertheless, I am obliged to Carol Barnes, Mark Bear, Ed Boyden, Emery Brown, Martha Constantine-Paton, Bob Desimone, Michale Fee, Guoping Feng, Mickey Goldberg, Alan Jasanoff, Yingxi Lin, Troy Littleton, Carlos Lois, Earl Miller, Peter Milner, Mortimer Mishkin, Chris Moore, Richard Morris, Morris Moscovitch, Ken Moya, Elisabeth Murray, Elly Nedivi, Russel Patterson, Tommy Poggio, Terry Sejnowski, Sebastian Seung, Mike Shadlen, Carla Shatz, Edie Sullivan, Mriganka Sur, Locky Taylor, Li-Huei Tsai, Chris Walsh, and Matt Wilson. I thank Leya Booth for transcribing these interviews quickly and accurately.

Helpful historical information came from conversations and emails with Brenda Milner, Bill Feindel, and Sandra McPherson at the Montreal Neurological Institute, and Marilyn Jonesgotman generously shared her 1977 interview with Henry. Alan Baddeley, Jean Gotman, Jake Kennedy, Ronald Lesser, Yvette Wong Penn, Arthur Reber, and Anthony Wagner contributed insights about cognitive and neural processes related to memory. Myriam Hyman imparted her advanced knowledge of ancient Greek, while Emilio Bizzi educated me about brain surgery, and Larry Squire advised me about terminology. Edie Sullivan helped me reconstruct the testing protocols we designed and carried out with Henry in the 1980s, and Mary Foley and Larry Wald helped document the activities of the epic night when Henry died.

For providing information about Hartford landmarks, I thank Brenda Miller, Manager of the Hartford History Center and Curator of the Hartford Collection at the Hartford Public Library, and Bill Faude, Project Historian at the Hartford History Center of the Hartford Public Library. At the MIT Science Library, Peter Norman was helpful in facilitating our research. I received a helpful analysis of Henry's memorable plane ride from Sandra Martin McDonough, pilot and flight instructor. Helen and Bob Sak and Gyorgy Buzsaki kindly sent me their reviews of the off-Broadway play about Henry.

For their help with the figures and photos, I am grateful to Henry's conservator Mr. M., Robert Ajemian, Jean Augustinack, Evelina Busa, Henry Hall, Sarah Holt, Producer for NOVA/PBS & Holt Productions, Bettiann McKay, Alex McWhinnie, Laura Pistorino, David Salat, André van der Kouwe, Victoria Vega, and Diana Woodruff-Pak.

Several people have been with me for the long haul and deserve special mention. Bettiann McKay is my administrative assistant and more importantly my friend and lifeline. Her contributions to my work would fill another book of this size. Suffice it to say that she was always there to give any kind of help whenever I needed it, and I will always be thankful for her generosity. John Growdon, my colleague for more than three decades, gave me sage advice, starting with my fledgling book proposal and continuing to the final draft of the book. Heaps of credit also go to Kathleen Lynch, a superb editor, who read every chapter more than once and gave me perceptive feedback as well as advice about all aspects of publishing.

Many friends offered encouragement. Over memorable dinners together, Lisa Scoville Dittrich helped me recapture our privileged childhood. My former students and postdocs gave their enthusiastic endorsement as did many others, including Edna Baginsky, Carol Christ, Holiday Smith Houck, David Margolis, Kerry Tribe, and Steve Pinker. Welcome inspiration also came from Susan Safford Andrews, Bobbi Topor Butler, Becky Crane Rafferty, Nancy Austin Reed, and Pat McEnroe Reno in Connecticut; from Doris, Jean-Claude, and Karine Welter in Paris; and from my fabulous Pier 7 neighbors in the Navy

Yard. Warmest thanks also to my Smith College classmates who are always an amazing source of support.

I am pleased to salute my fellow faculty members in MIT's Department of Brain and Cognitive Sciences. I have benefitted greatly from decades of interactions with them and am excited and inspired by their extraordinary work. I also want to acknowledge the wonderful graduate students and postdocs in our department who responded quickly and cheerfully to my emails asking for miscellaneous information that had nothing to do with science.

An affectionate thank you goes to my children, Zachary Corkin, Jocelyn Corkin Mortimer, and Damon Corkin for their love, encouragement, and praise—and for keeping me humble. They and their families are a source of energy and delight. One of the great joys of writing this book was the discovery that Jocelyn is a superb editor. She meticulously read many drafts and caught countless errors that others had missed. Her contributions improved the telling of Henry's story by many orders of magnitude, and I thank her most sincerely. I also greatly appreciate the endless interest and enthusiasm of other family members—Jane Corkin, Donald Corkin, and Patricia and Jake Kennedy and their family.

I am fortunate to have the guidance of the Wyley Agency in achieving my dream to write this book. The members of their highly professional and gifted staff do an impressive job in carrying out their various responsibilities. In particular, I want to thank Andrew Wylie, Scott Moyers, Rebecca Nagel, and Kristina Moore, who are exceptional people with whom to work.

At Perseus Books, I received much needed editorial and production help from Lara Heimert, Ben Reynolds, Chris Granville, Katy O'Donnell, and Rachel King. I am grateful for their acumen and patience, and for their willingness to become immersed in the life of Henry Molaison and the neuroscience of memory.

Notes

Prologue

1. Neuroscience is a giant tent that covers diverse disciplines, all intended to advance knowledge about the brain and nervous system. Systems neuroscience is a branch of neuroscience whose mission is to describe the specialization of distinct circuits of interconnected neurons that give rise to specific kinds of behavior, such as declarative and nondeclarative memory. The systems include sensory capacities such as vision, hearing, and touch, and high-order processes such as problem solving, goal-directed behavior, spatial ability, motor control, and language. Studying Henry gave us the extraordinary opportunity to propel the science of human memory forward by examining processes distributed throughout the brain; W. B. Scoville and B. Milner, "Loss of Recent Memory after Bilateral Hippocampal Lesions," *Journal of Neurology, Neurosurgery, and Psychiatry* 20 (1957): 11–21.

2. Scoville and Milner, "Loss of Recent Memory after Bilateral Hippocampal Lesions."

3. Ibid. In previous memory testing with Henry, Milner had used test materials presented through vision and hearing.

4. P. J. Hilts, "A Brain Unit Seen as Index for Recalling Memories," *New York Times* (1991, September 24); P. J. Hilts, *Memory's Ghost: The Strange Tale of Mr. M. and the Nature of Memory* (New York: Simon & Schuster, 1995).

5. N. J. Cohen and L. R. Squire, "Preserved Learning and Retention of Pattern-Analyzing Skill in Amnesia: Dissociation of Knowing How and Knowing That," *Science* 210 (1980): 207–10.

Chapter One: Prelude to Tragedy

1. O. Temkin, *The Falling Sickness: A History of Epilepsy from the Greeks to the Beginnings of Modern Neurology* (Baltimore, MD: Johns Hopkins Press, 1971).

2. Ibid.

3. Ibid.

4. W. Feindel et al., "Epilepsy Surgery: Historical Highlights 1909–2009," *Epilepsia* 50 (2009): 131–51.

5. Ibid.

6. M. D. Niedermeyer et al., "Rett Syndrome and the Electroencephalogram," *American Journal of Medical Genetics* 25 (2005): 1096–8628; H. Berger, "Über Das Elektrenkephalogramm Des Menschen " *European Archives of Psychiatry and Clinical Neuroscience* 87 (1929): 527–70.

7. W. Feindel et al., "Epilepsy Surgery: Historical Highlights 1909-2009," *Epilepsia* 50 (2009): 131–51; W. B. Scoville et al., "Observations on Medial Temporal Lobotomy and Uncotomy in the Treatment of Psychotic States; Preliminary Review of 19 Operative Cases Compared with 60 Frontal Lobotomy and Undercutting Cases," *Proceedings for the Association for Research in Nervous and Mental Disorders* 31 (1953): 347–73; O. Temkin, *The Falling Sickness: A History of Epilepsy from the Greeks to the Beginnings of Modern Neurology* (Baltimore, MD: Johns Hopkins Press, 1971); B. V. White et al., *Stanley Cobb: A Builder of the Modern Neurosciences* (Charlottesville, VA: University Press of Virginia, 1984).

8. W. Feindel et al., "Epilepsy Surgery: Historical Highlights 1909–2009," *Epilepsia* 50 (2009): 131–51.

9. Jack Quinlan, October 8, 1945.

10. W. B. Scoville, "Innovations and Perspectives," *Surgical Neurology* 4 (1975): 528.

11. W. B. Scoville and B. Milner, "Loss of Recent Memory after Bilateral Hippocampal Lesions," *Journal of Neurology, Neurosurgery, and Psychiatry* 20 (1957): 11–21.

12. Liselotte K. Fischer, Unpublished report of psychological testing, Hartford Hospital, August 24, 1953.

Chapter Two: "A Frankly Experimental Operation"

1. J. El-Hai, *The Lobotomist: A Maverick Medical Genius and His Tragic Quest to Rid the World of Mental Illness* (Hoboken, NJ: J. Wiley, 2005); John F. Kennedy Memorial Library, "The Kennedy Family: Rosemary Kennedy"; www.jfklibrary.org/JFK/The-Kennedy-Family/Rosemary-Kennedy.aspx (accessed November 2012).

2. J. L. Stone, "Dr. Gottlieb Burckhardt—The Pioneer of Psychosurgery," *Journal of the History of the Neurosciences* 10 (2001): 79–92; El-Hai, *The Lobotomist*.

3. B. Ljunggren et al., "Ludvig Puusepp and the Birth of Neurosurgery in Russia," *Neurosurgery Quarterly* 8 (1998): 232–35.

4. C. F. Jacobsen et al., "An Experimental Analysis of the Functions of the Frontal Association Areas in Primates," *Journal of Nervous and Mental Disorders* 82 (1935): 1–14.

5. E. Moniz, *Tentatives Opératoires dans le Traitement de Certaines Psychoses* (Paris, France: Masson, 1936).

6. Ibid.

7. E. Moniz, "Prefrontal Leucotomy in the Treatment of Mental Disorders," *American Journal of Psychiatry* 93 (1937): 1379–85; El-Hai, *The Lobotomist*.

8. W. Freeman and J. W. Watts, *Psychosurgery in the Treatment of Mental Disorders and Intractable Pain* (Springfield, IL: C. C. Thomas, 1950); J. D. Pressman, *Last Resort: Psychosurgery and the Limits of Medicine* (Cambridge Studies in the History of Medicine) (New York: Cambridge University Press, 1998); El-Hai, *The Lobotomist*.

9. D. G. Stewart and K. L. Davis, "The Lobotomist," *American Journal of Psychiatry* 165 (2008): 457–58; El-Hai, *The Lobotomist*.

10. J. E. Rodgers, *Psychosurgery: Damaging the Brain to Save the Mind* (New York: HarperCollins, 1992); El-Hai, *The Lobotomist*.

11. Pressman, *Last Resort*; El-Hai, *The Lobotomist*.

12. Pressman, *Last Resort*.

13. National Commission for the Protection of Human Subjects of Biomedical and Behavioral Research, *Psychosurgery: Report and Recommendations* (Washington, DC: DHEW Publication No. [OS] 77–0001,1977); available online at videocast.nih.gov/pdf/ohrp_psychosurgery.pdf (accessed November 2012).

14. W. B. Scoville et al., "Observations on Medial Temporal Lobotomy and Uncotomy in the Treatment of Psychotic States: Preliminary Review of 19 Operative Cases Compared with 60 Frontal Lobotomy and Undercutting Cases," Proceedings for the Association for Research in Nervous and Mental Disorders 31 (1953): 347–73.

15. W. Penfield and M. Baldwin, "Temporal Lobe Seizures and the Technic of Subtotal Temporal Lobectomy," *Annals of Surgery* 136 (1952): 625–34, available online at www.ncbi.nlm.nih.gov/pmc/articles/PMC1803045/pdf/annsurg01421–0076.pdf (accessed November 2012); Scoville et al., "Observations on Medial Temporal Lobotomy and Uncotomy."

16. W. B. Scoville and B. Milner, "Loss of Recent Memory after Bilateral Hippocampal Lesions," *Journal of Neurology, Neurosurgery, and Psychiatry* 20 (1957): 11–21, available online at jnnp.bmj.com/content/20/1/11.short (accessed November 2012).

17. Ibid.

18. MacLean, "Some Psychiatric Implications"; Scoville and Milner, "Loss of Recent Memory."

19. Scoville and Milner, "Loss of Recent Memory"; S. Corkin et al., "H.M.'s Medial Temporal Lobe Lesion: Findings from MRI," *Journal of Neuroscience* 17 (1997): 3964–79.

20. P. Andersen et al., *Historical Perspective: Proposed Functions, Biological Characteristics, and Neurobiological Models of the Hippocampus* (New York: Oxford University Press, 2007); J. W. Papez, "A Proposed Mechanism of Emotion. 1937," *Journal of Neuropsychiatry and Clinical Neurosciences* 7 (1995): 103–12; MacLean, "Some Psychiatric Implications."

21. Scoville and Milner, "Loss of Recent Memory."

Chapter Three: Penfield and Milner

1. W. Penfield and B. Milner, "Memory Deficit Produced by Bilateral Lesions in the Hippocampal Zone," *AMA Arch Neurol Psychiatry* 79:5 (May 1958) 475–97; B. Milner, "The Memory Defect in Bilateral Hippocampal Lesions," *Psychiatric Research Reports of the American Psychiatric Association* 11 (1959): 43–58.

2. W. Penfield, *No Man Alone: A Neurosurgeon's Life* (Boston, MA: Little, Brown, 1977).

3. W. Penfield, "Oligodendroglia and Its Relation to Classical Neuroglia," *Brain* 47 (1924): 430–52.

4. O. Foerster and W. Penfield, "The Structural Basis of Traumatic Epilepsy and Results of Radical Operation," *Brain* 53 (1930): 99–119.

5. W. Penfield and M. Baldwin, "Temporal Lobe Seizures and the Technic of Subtotal Temporal Lobectomy," *Annals of Surgery* 136 (1952): 625–34, available online at www.ncbi.nlm.nih.gov/pmc/articles/PMC1803045/pdf /annsurg01421–0076.pdf (accessed November 2012); P. Robb, *The Development of Neurology at McGill* (Montreal: Osler Library, McGill University, 1989); W. Feindel et al., "Epilepsy Surgery: Historical Highlights 1909–2009," *Epilepsia* 50 (2009): 131–51.

6. F. C. Bartlett, *Remembering: A Study in Experimental and Social Psychology.* (New York: Cambridge University Press, 1932); C.W.M. Whitty and O. L. Zangwill, *Amnesia* (London: Butterworths, 1966).

7. Penfield and Milner, "Memory Deficit Produced by Bilateral Lesions in the Hippocampal Zone"; Milner, "The Memory Deficit Bilateral Hippocampal Lesions."

8. Ibid; W. Penfield and H. Jasper, *Epilepsy and the Functional Anatomy of the Human Brain* (Boston: Little, Brown, 1954).

9. W. Penfield and G. Mathieson, "Memory: Autopsy Findings and Comments on the Role of Hippocampus in Experiential Recall," *Archives of Neurology* 31 (1974): 145–54.

10. S. Demeter et al., "Interhemispheric Pathways of the Hippocampal Formation, Presubiculum, and Entorhinal and Posterior Parahippocampal Cortices in the Rhesus Monkey: The Structure and Organization of the Hippocampal Commissures," *Journal of Comparative Neurology* 233 (1985): 30–47.

11. Penfield and Milner, "Memory Deficit Produced by Bilateral Lesions in the Hippocampal Zone"; Milner, "The Memory Deficit Bilateral Hippocampal Lesions."

12. B. Milner and W. Penfield, "The Effect of Hippocampal Lesions on Recent Memory," *Transactions of the American Neurological Association* (1955–1956): 42–48; W. B. Scoville and B. Milner, "Loss of Recent Memory after Bilateral Hippocampal Lesions," *Journal of Neurology, Neurosurgery, and Psychiatry* 20 (1957): 11–21, available online at jnnp.bmj.com/content/20/1/11.short (accessed November 2012).

13. Scoville and Milner, "Loss of Recent Memory."

14. W. B. Scoville, "The Limbic Lobe in Man," *Journal of Neurosurgery* 11 (1954): 64–66; Scoville and Milner, 1957.

15. Scoville and Milner, "Loss of Recent Memory"; B. Milner, "Psychological Defects Produced by Temporal Lobe Excision," *Research Publications—Association for Research in Nervous and Mental Disease* 36 (1958): 244–57.

16. Scoville and Milner, "Loss of Recent Memory."

17. W. B. Scoville, "Amnesia after Bilateral Medial Temporal-Lobe Excision: Introduction to Case H.M.," *Neuropsychologia* 6 (1968): 211–13; W. B. Scoville, "Innovations and Perspectives," *Surgical Neurology* 4 (1975): 528–30; L. Dittrich, "The Brain that Changed Everything," *Esquire* 154 (November 2010): 112–68.

18. B. Milner, "Intellectual Function of the Temporal Lobes," *Psychological Bulletin* 51 (1954): 42–62.

19. W. Penfield and E. Boldrey, "Somatic Motor and Sensory Representation in the Cerebral Cortex of Man as Studied by Electrical Stimulation," *Brain* 60 (1937): 389–443; W. Feindel and W. Penfield, "Localization of Discharge in Temporal Lobe Automatism," *Archives of Neurology & Psychiatry* 72 (1954): 605–30; W. Penfield and L. Roberts, *Speech and Brain-Mechanisms* (Princeton, NJ: Princeton University Press, 1959).

20. S. Corkin, "Tactually-Guided Maze Learning in Man: Effects of Unilateral Cortical Excisions and Bilateral Hippocampal Lesions," *Neuropsychologia* 3 (1965): 339–51, available online at web.mit.edu/bnl/pdf/Corkin_1965.pdf (accessed November 2012).

Chapter Four: Thirty Seconds

1. D. O. Hebb, *The Organization of Behavior: A Neuropsychological Theory* (New York: Wiley, 1949).

2. S. R. Cajal, "La Fine Structure des Centres Nerveux," *Proceedings of the Royal Society of London* 55 (1894): 444–68.

3. C. J. Shatz, "The Developing Brain," *Scientific American* 267 (1992): 60–67; available online at cognitrn.psych.indiana.edu/busey/q551/PDFs/Mind BrainCh2.pdf (accessed November 2012).

4. E. R. Kandel, "The Molecular Biology of Memory Storage: A Dialogue between Genes and Synapses," *Science* 294 (2001): 1030–38; Kandel, *In Search of Memory*.

5. Hebb, *The Organization of Behavior*; Kandel, *In Search of Memory*.

6. L. Prisko, *Short-Term Memory in Focal Cerebral Damage* (unpublished dissertation; Montreal: McGill University, 1963).

7. E. K. Warrington et al., "The Anatomical Localisation of Selective Impairment of Auditory Verbal Short-Term Memory," *Neuropsychologia* 9 (1971): 377–87.

8. Ibid.

9. N. Kanwisher, "Functional Specificity in the Human Brain: A Window into the Functional Architecture of the Mind," *Proceedings of the National Academy of Sciences of the United States of America* 107 (2010): 11163–70.

10. E. K. Miller and J. D. Cohen, "An Integrative Theory of Prefrontal Cortex Function," *Annual Review of Neuroscience* 24 (2001): 167–202; available online at web.mit.edu/ekmiller/Public/www/miller/Publications/Miller_Cohen _2001.pdf (accessed November 2012).

11. B. Milner, "Reflecting on the Field of Brain and Memory," Lecture of November 18, 2008 (Washington, DC: Society for Neuroscience).

12. J. Brown, "Some Tests of the Decay Theory of Immediate Memory," *Quarterly Journal of Experimental Psychology* 10 (1958): 12–21.

13. L. R. Peterson and M. J. Peterson, "Short-Term Retention of Individual Verbal Items," *Journal of Experimental Psychology* 58 (1959): 193–98; available online at hs-psychology.ism-online.org/files/2012/08/Peterson-Peterson-1959-duration-of-STM.pdf (accessed November 2012).

14. S. Corkin, "Some Relationships between Global Amnesias and the Memory Impairments in Alzheimer's Disease," in *Alzheimer's Disease: A Report of Progress in Research*, ed. S. Corkin et al. (New York: Raven Press, 1982), 149–64.

15. B. Milner et al., "Further Analysis of the Hippocampal Amnesic Syndrome: 14-Year Follow-up Study of H.M.," *Neuropsychologia* 6 (1968): 215–34.

16. B. Milner, "Effects of Different Brain Lesions on Card Sorting: The Role of the Frontal Lobes," *Archives of Neurology* 9 (1963): 100–10.

17. A. Jeneson and L. R. Squire, "Working Memory, Long-Term Memory, and Medial Temporal Lobe Function," *Learning & Memory* 19 (2012): 15–25.

18. N. Wiener, *Cybernetics: or, Control and Communication in the Animal and the Machine* (Cambridge: MIT Press, 1948).

19. G. A. Miller et al., *Plans and the Structure of Behavior* (New York: Holt, 1960).

20. R. C. Atkinson and R. M. Shiffrin, "Human Memory: A Proposed System and Its Control Processes," in *The Psychology of Learning and Motivation: Advances in Research and Theory*, vol. 2, ed. K. W. Spence and J. T. Spence (New York: Academic Press, 1968), 89–195; available online at tinyurl.com/aa4w696 (accessed November 2012).

21. A. D. Baddeley and G.J.L. Hitch, "Working Memory," in *The Psychology of Learning and Motivation: Advances in Research and Theory*, ed. G. H. Bower (New York: Academic Press, 1974), 47–89.

22. B. R. Postle, "Working Memory as an Emergent Property of the Mind and Brain," *Neuroscience* 139 (2006): 23–38; M. D'Esposito, "From Cognitive to Neural Models of Working Memory," *Philosophical Transactions of the Royal Society of London, Series B: Biological Sciences* 362 (2007): 761–72; J. Jonides et al., "The Mind and Brain of Short-Term Memory," *Annual Review of Psychology* 59 (2008): 193–224.

23. Miller and Cohen, "An Integrative Theory of Prefrontal Cortex Function," Annual Review of Neuroscience 24 (2001): 167–202.

24. Ibid.

Chapter Five: Memories Are Made of This

1. Scoville's notes and sketches were the basis of a set of detailed drawings by another neurosurgeon, Lamar Roberts, which accompanied Scoville and Milner's 1957 paper.

2. P. C. Lauterbur, "Image Formation by Induced Local Interactions: Examples of Employing Nuclear Magnetic Resonance," *Nature* 242 (1973): 1901; P. Mansfield and P.K. Grannell, "NMR 'Diffraction' in Solids?," *Journal of Physics C: Solid State Physics* 6 (1973): L422.

3. S. Corkin et al., "H.M.'s Medial Temporal Lobe Lesion: Findings from MRI," *Journal of Neuroscience* 17 (1997): 3964–79.

4. H. Eichenbaum, *The Cognitive Neuroscience of Memory: An Introduction* (New York: Oxford University Press, 2011).

5. B. Milner et al., "Further Analysis of the Hippocampal Amnesic Syndrome: 14-Year Follow-up Study of H.M.," *Neuropsychologia* 6 (1968): 215–34.

6. Ibid.

7. Corkin, "H.M.'s Medial Temporal Lobe Lesion."

8. H. Eichenbaum et al., "Selective Olfactory Deficits in Case H.M.," *Brain* 106 (1983): 459–72.

9. Ibid.

10. Ibid.

11. Ibid.

12. Henry's impairment on the route-finding task, performed as a laboratory experiment, reinforced the theory introduced in John O'Keefe and Lynn Nadel's classic 1978 book, *The Hippocampus as a Cognitive Map* (New York: Oxford University Press), which combined information from theoretical, behavioral, anatomical, and physiological sources to propose that the hippocampus oversees cognitive mapping and memory for spatial layouts and experiences moving in space.

13. B. Milner, "Visually-Guided Maze Learning in Man: Effects of Bilateral Hippocampal, Bilateral Frontal, and Unilateral Cerebral Lesions," *Neuropsychologia* 3 (1965): 317–38.

14. S. Corkin, "Tactually-Guided Maze Learning in Man: Effects of Unilateral Cortical Excisions and Bilateral Hippocampal Lesions," *Neuropsychologia* 3 (1965): 339–51.

15. S. Corkin, "What's New with the Amnesic Patient H.M.?," *Nature Reviews Neuroscience* 3 (2002): 153–60.

16. S. Corkin et al., "H.M.'s Medial Temporal Lobe Lesion."

17. V. D. Bohbot and S. Corkin, "Posterior Parahippocampal Place Learning in H.M.," *Hippocampus* 17 (2007): 863–72.

18. Ibid.

Chapter Six: "An Argument with Myself"

1. J. D. Payne, "Learning, Memory, and Sleep in Humans," *Sleep Medicine Clinics* 6 (2011): 15–30; R. Stickgold and M. Tucker, "Sleep and Memory: In Search of Functionality," in *Augmenting Cognition*, eds, I. Segev et al. (Boca Raton, FL: CRC Press, 2011); 83–102.

2. P. Broca, "Sur la Circonvolution Limbique et la Scissure Limbique," *Bulletins de la Société d'Anthropologie de Paris* 12 (1877): 646–57; J. W. Papez, "A Proposed Mechanism of Emotion," *Archives of Neurology and Psychiatry* 38 (1937): 725–43.

3. Papez, "A Proposed Mechanism of Emotion"; J. Nolte and J. W. Sundsten, *The Human Brain: An Introduction to Its Functional Anatomy* (Philadelphia, PA: Mosby, 2009); K. A. Lindquist et al., "The Brain Basis of Emotion: A Meta-Analytic Review," *Behavioral and Brain Sciences* 35 (2012): 121–43.

4. P. Ekman, "Basic Emotions," in *Handbook of Cognition and Emotion*, eds, T. Dalgleish et al. (New York: Wiley, 1999), 45–60.

5. E. A. Kensinger and S. Corkin, "Memory Enhancement for Emotional Words: Are Emotional Words More Vividly Remembered Than Neutral Words?," *Memory and Cognition* 31 (2003): 1169–80; E. A. Kensinger and

S. Corkin, "Two Routes to Emotional Memory: Distinct Neural Processes for Valence and Arousal," *Proceedings of the National Academy of Sciences* 101 (2004): 3310–5.

Chapter Seven: Encode, Store, Retrieve

1. C. E. Shannon, "A Mathematical Theory of Communication," *Bell System Technical Journal* 27 (1948): 379–423, 623–56; G. A. Miller, "The Magical Number Seven, Plus or Minus Two: Some Limits on Our Capacity for Processing Information," *Psychological Review* 63 (1956): 81–97.

2. A. S. Reber, "Implicit Learning of Artificial Grammars 1," *Journal of Verbal Learning and Verbal Behavior* 6 (1967): 855–63; N. J. Cohen and L. R. Squire, "Preserved Learning and Retention of Pattern-Analyzing Skill in Amnesia: Dissociation of Knowing How and Knowing That," *Science* 210 (1980): 207–10; L. R. Squire and S. Zola-Morgan, "Memory: Brain Systems and Behavior," *Trends in Neuroscience* 11 (1988): 170–5.

3. F.I.M. Craik and R. S. Lockhart, "Levels of Processing: A Framework for Memory Research," *Journal of Verbal Learning and Verbal Behavior* 11 (1972): 671–84; F.I.M. Craik and E. Tulving, "Depth of Processing and the Retention of Words in Episodic Memory," *Journal of Experimental Psychology* 104 (1975): 268–94.

4. Ibid.

5. S. Corkin, "Some Relationships between Global Amnesias and the Memory Impairments in Alzheimer's Disease," in *Alzheimer's Disease: A Report of Progress in Research*, eds, S. Corkin et al. (New York: Raven Press, 1982), 149–64.

6. Ibid.

7. Corkin, "Some Relationships"; K. Velanova et al., "Evidence for Frontally Mediated Controlled Processing Differences in Older Adults," *Cerebral Cortex* 17 (2007): 1033–46.

8. R. L. Buckner and J. M. Logan, "Frontal Contributions to Episodic Memory Encoding in the Young and Elderly," in *The Cognitive Neuroscience of Memory*, eds, A. Parker et al. (New York: Psychology Press, 2002), 59–81; U. Wagner et al., "Effects of Cortisol Suppression on Sleep-Associated Consolidation of Neutral and Emotional Memory," *Biological Psychiatry* 58 (2005): 885–93.

9. J. A. Ogden, *Trouble in Mind: Stories from a Neuropsychologist's Casebook* (New York: Oxford University Press, 2012).

10. J. D. Spence, *The Memory Palace of Matteo Ricci* (London: Quercus, 1978).

11. A. Raz et al., "A Slice of Pi: An Exploratory Neuroimaging Study of Digit Encoding and Retrieval in a Superior Memorist," *Neurocase* 15 (2009): 361–72.

12. Raz, "A Slice of Pi"; K. A. Ericsson, "Exceptional Memorizers: Made, Not Born," *Trends in Cognitive Science* 7 (2003): 233–5.

13. Buckner and Logan, "Frontal Contributions to Episodic Memory Encoding."

14. H. A. Lechner et al., "100 Years of Consolidation—Remembering Müller and Pilzecker, " *Learning Memory* 6 (1999): 77–87.

15. Ibid.

16. Ibid.

17. C. P. Duncan, "The Retroactive Effect of Electroshock on Learning," *Journal of Comparative Psychology* 42 (1949): 32–44; J. L. McGauch, "Memory—A Century of Consolidation," *Science* 287 (2000): 248–51; S. J. Sara and B. Hars, "In Memory of Consolidation," *Learning and Memory* 13 (2006): 515–21.

18. H. Eichenbaum, "Hippocampus: Cognitive Processes and Neural Representations That Underlie Declarative Memory," *Neuron* 44 (2004): 109–20.

19. Eichenbaum, "Hippocampus"; D. Shohamy and A. D. Wagner, "Integrating Memories in the Human Brain: Hippocampal-Midbrain Encoding of Overlapping Events," *Neuron* 60 (2008): 378–89.

20. W. B. Scoville and B. Milner, "Loss of Recent Memory after Bilateral Hippocampal Lesions," *Journal of Neurology, Neurosurgery, and Psychiatry* 20 (1957): 11–21; B. Milner, "Psychological Defects Produced by Temporal Lobe Excision," *Research Publications—Association for Research in Nervous and Mental Disease* 36 (1958): 244–57.

21. Ibid.; W. Penfield and B. Milner, "Memory Deficit Produced by Bilateral Lesians in the Hippocampal Zone," A.M.A. Archives of *Neurology & Psychiatry* 79 (1950): 475–97. To examine the intricacies of the cognitive and neural processes that support each stage of memory in humans, neuroscientists have turned to experiments with a variety of animal species. These investigations have documented memory formation at several levels—improved memory performance, increase or decrease in the firing rate of neurons, and structural and functional modifications in cells and molecules. These alterations are all evidence of *neural plasticity*, the brain's ability to change as a result of experience. The eventual goal of this ongoing research is to integrate knowledge from all levels to create a comprehensive description of how learning and memory come about.

Monkeys are animals of choice for insights about cognitive processes that are similar to those in humans. They can learn more complex tasks than rodents, especially when it comes to cognitive flexibility—the ability to set goals and then execute the thoughts and actions to achieve them. But monkeys are expensive to house and require months of training because the cognitive tasks researchers want them to perform are so complex. As a result, rates and mice are widely used for memory research. Each species has its advantages. Mice are the ideal subjects when genetic models or manipulations are needed.

The seeds of gene targeting were sown in 1977, and the technology evolved to the point that it is now used in thousands of laboratories worldwide. In 2007, the Nobel Prize in Physiology or Medicine was awarded to Mario Capecchi, Sir Martin Evans, and Oliver Smithies "for their discoveries of principles for introducing specific gene modifications in mice by the use of embryonic stem cells." This method can be used to knock out function in specific tissues in the mouse and in doing so replicate hundreds of human diseases. The advantage of mouse models is that they allow scientists to study diseases with greater precision than is possible in humans, with the hope of creating new therapies targeted to the underlying pathology. See Gene Targeting 1977–Present. Nobel Prize Lecture http://www.nobelprize.org:nobel_prizes:medicine:laureates:2007:capecchi -lecture.html.

Gene targeting in rats has not been possible until recently, but rats have been much more fully characterized in the laboratory in terms of their anatomy, physiology, and behavior, and their larger brains make recording from neurons in active animals easier. Much of neuroscience research uses both species in a complementary fashion to address unanswered questions. An interesting array of other species has been used for more specialized purposes—Zebra finches for the study of song learning, ferrets for their superb visual system, and even marine slugs, known as *aplysia*, for their enormous and easily accessible neurons.

The history of memory research is a synthesis of memory experiments from multiple species, each of which contributes critical advances along the way. Although countless questions remain unanswered, the past few decades have brought a dizzying amount of knowledge suggesting how a learning experience is transformed into lasting changes in brain circuits.

22. McGauch, "Memory—A Century of Consolidation"; Memory consolidation theorists speculated that long-term declarative memory, Henry's nemesis, depends on such close interaction and coordination between the workings of the hippocampus and processes in the cortex. Henry's intact cerebral cortex could not do the job by itself. In 2012, research cintnues to focus on how the hippocampal system interacts with cortical circuits to consolidate and store memories. Because consolidation takes place gradually, it is reasonable to suppose that multiple mechanisms in the hippocampus and cortex are recruited along the way. See D. Marr, "Simple Memory: A Theory for Archicortex," *Philosophical Transactions of the Royal Society of London, Series B, Biological* Sciences 262 (1971): 23–81; L. R. Squire et al., "The Medial Temporal Region and Memory Consolidation: A New Hypothesis," in *Memory Consolidation:*

Psychobiology of Cognition, eds, H. Weingartner et al. (Hillsdale, NJ: Lawrence Erlbaum Associates, 1984), 185–210; and J. L. McClelland et al., "Why There Are Complementary Learning Systems in the Hippocampus and Neocortex: Insights from the Successes and Failures of Connectionist Models of Learning and Memory," *Psychological Review* 102 (1995): 419–57.

23. S. Ramón y Cajal, "La Fine Structure des Centres Nerveux," *Proceedings of the Royal Society of London* 55 (1894): 444–68; D. O. Hebb, *The Organization of Behavior: A Neuropsychological Theory* (New York: John Wiley & Sons, 1949).

24. T. Lømo, "Frequency Potentiation of Excitatory Synaptic Activity in the Dentate Areas of the Hippocampal Formation," *Acta Physiologica Scandinavica* 68 (1966): 128; T.V.P. Bliss and T. Lømo, Long-Lasting Potentiation of Synaptic Transmission in the Dentate Area of the Anaesthetized Rabbit Following Stimulation of the Perforant Path," *Journal of Physiology* 232 (1973): 331–56; R. M. Douglas and G. Goddard, "Long-Term Potentiation of the Perforant Path-Granule Cell Synapse in the Rat Hippocampus," *Brain Research* 86 (1975): 205–15.

25. S. J. Martin et al., "Synaptic Plasticity and Memory: An Evaluation of the Hypothesis," *Annual Review of Neuroscience* 23 (2000): 649–711; T. Bliss et al., "Synaptic Plasticity in the Hippocampus," in *The Hippocampus Book*, eds, P. Anderson et al. (New York: Oxford University Press, 2007), 343–474.

26. Ibid.

27. Next, these scientists asked whether this learning deficit applied to all kinds of learning or was specific to spatial learning. He trained rats on a simple visual-discrimination task where they were allowed to choose between two platforms based on how they looked—a grey one that floated and provided escape, and a black-and-white striped one that sank. This task did not require spatial learning. Rats who received the drug to block LTP performed the visual-discrimination task normally, indicating that the hippocampus was not necessary for this task. The sharp contrast between the dramatic deficit in spatial (declarative) learning and the intact discrimination (nondeclarative) learning is reminiscent of Henry's postoperative inability to find his way to the bathroom in the hospital alongside his facility in learning new motor skills. See R. G. Morris et al., "Selective Impairment of Learning and Blockade of Long-Term Potentiation by an N-Methyl-D-Aspartate Receptor Antagonist, Ap5," *Nature* 319 (1986): 774–6.

28. J. Z. Tsien, et al., "Subregion-and Cell Type-Restricted Gene Knockout in Mouse Brain," *Cell* 87 (1996): 1317–26; T. J. McHugh, et al., "Impaired Hippocampal Representation of Space in CA1-Specific NMDAR1 Knockout Mice," *Cell* 87 (1996): 1339–49; A. Rotenberg, et al., "Mice Expressing Activated CaMKII Lack Low Frequency LTP and Do Not Form Stable Place Cells in the CA1 Region of the Hippocampus," *Cell* 87 (1996): 1351–61.

29. T.V.P. Bliss and S. F. Cooke, "Long-Term Potentiation and Long-Term Depression: A Clinical Perspective," *Clinics* 66 (2011): 3–17.

30. J. O'Keefe and J. Dostrovsky, "The Hippocampus as a Spatial Map: Preliminary Evidence from Unit Activity in the Freely-Moving Rat," *Brain Research* 34 (1971): 171–5.

31. Y. L. Qin et al. "Memory Reprocessing in Corticocortical and Hippocampocortical Neuronal Ensembles," *Philosophical Transactions of the Royal Society of London, Series B, Biological Sciences* 352 (1997): 1525–33.

32. J. D. Payne, Learning, Memory, and Sleep in Humans," *Sleep Medicine Clinics* 6 (2011): 145–56.

33. K. Louie and M. A. Wilson, "Temporally Structured Replay of Awake Hippocampal Ensemble Activity During Rapid Eye Movement Sleep," *Neuron* 29 (2001): 145–56.

34. Ibid.

35. A. K. Lee and M. A. Wilson, "Memory of Sequential Experience in the Hippocampus During Slow Wave Sleep," *Neuron* 36 (2002): 1183–94. Memory replay in *awake* rats also advances our understanding of consolidation. In 2006, Wilson and colleagues discovered that after a rat ran a novel track and then stopped to take time out to groom, whisk its whiskers, or just stand still, the memories of locations in the maze formed in its hippocampus were played back in reverse order—the place cells associated with the end of the track fired first, and those related to the beginning fired last. This backward instant replay suggests that the rat stopped to literally think back in time, contemplating, assimilating, and consolidating what it had just experienced. As two neuroscientists at Rutgers University showed in 2007, awake rats can also replay sequences forward—in the same order in which they were experienced. The puzzle is: What are these rats thinking about, and why do they engage in replay? If this is not full-blown thought, then it is at least a gigantic leap in that direction. See D. J. Foster and M. A. Wilson, "Reverse Replay of Behavioural Sequences in Hippocampal Place Cells During the Awake State," *Nature* 440 (2006): 680–3

36. Ibid.; and K. Diba and G. Buzsaki, "Forward and Reverse Hippocampal Place-Cell Sequences During Ripples," *Nature Neuroscience* 10m (2007): 1241–2.

37. D. Ji and M.A. Wilson, Coordinated Memory Replay in the Visual Cortex and Hippocampus During Sleep," *Nature Neuroscience* 10 (2007): 100–7.

38. E. Tulving and D. M. Thomson, "Encoding Specificity and Retrieval Processes in Episodic Memory," *Psychological Review* 80 (1973): 352–73.

39. H. Schmolck, et al., "Memory Distortions Develop over Time: Recollections of the O.J. Simpson Trial Verdict after 15 and 32 Months," *Psychological Science* 11 (2000): 39–45.

40. J. Przybyslawski and S. J. Sara, "Reconsolidation of Memory after Its Reactivation," *Behavioural Brain Research* 84(1997): 241–6.

41. Ibid.

42. O. Hardt et al., "A Bridge over Troubled Water: Reconsolidation as a Link between Cognitive and Neuroscientific Memory Research Traditions," *Annual Review of Psychology* 61 (2010): 141–67; See also D. Schiller et al., "Preventing the Return of Fear in Humans Using Reconsolidation Update Mechanisms," *Nature* 463 (2010): 49–53.

43. J. T. Wixted, "The Psychology and Neuroscience of Forgetting," *Annual Review of Psychology* 55 (2004): 235–69.

44. D. M. Freed et al., "Forgetting in H.M.: A Second Look," *Neuropsychologia* 25 (1987): 461–71.

45. Freed, "Forgetting in H.M."; D. M. Freed and S. Corkin, "Rate of Forgetting in H.M.: 6-Month Recognition," *Behavioral Neuroscience* 102 (1988): 823–7.

46. R. C. Atkinson and J. F. Juola, "Search and Decision Processes in Recognition Memory," in *Contemporary Developments in Mathematical Psychology: Learning, Memory, and Thinking*, eds, D. H. Krantz (San Francisco, CA: W. H. Freeman, 1974), 242–93; G. Mandler, "Recognizing: The Judgement of Previous Occurrence," *Psychological Review* 87 (1980): 252–71; L. L. Jacoby, "A Process Dissociation Framework: Separating Automatic from Intentional Uses of Memory," *Journal of Memory and Language* 30 (1991): 513–41.

47. J. P. Aggleton and M. W. Brown, "Episodic Memory, Amnesia, and the Hippocampal-Anterior Thalamic Axis," *Behavioral and Brain Science* 22 (1999): 425–44.

48. Freed, "Forgetting in H.M."; Freed and Corkin, "Rate of Forgetting in H.M."; and Aggleton and brown, "Episodic Memory."

49. C. Ranganath et al., "Dissociable Correlates of Recollection and Familiarity within the Medial Temporal Lobes," *Neuropsychologia* 42 (203): 2–13.

50. Ibid.

51. Ibid.

52. B. Bowles et al., "Impaired Familiarity with Preserved Recollection after Anterior Temporal-Lobe Resection That Spares the Hippocampus," *Proceedings of the National Academy of Sciences* 104 (2007): 16382–7; M. W. Brown et al., "Recognition Memory: Material, Processes, and Substrates: *Hippocampus* 20 (2010): 1228–44. In 2011, cognitive neuroscientists at New York University proposed a different view about the organization of recognition memory in medial temporal-lobe areas. Their functional MRI results in healthy research participants suggested that the perirhinal cortex was specialized for imagining individual objects, whereas the parahippocampal cortex was specialized for imagining scenes. See B. P. Staresina et al., "Perirhinal and Parahippocamal Cortices Differentially Contribute to Later Recollection of Object- and Scene-Related Event Details," *Journal of Neuroscience* 31 (2011): 8739–47.

Chapter Eight: Memory without Remembering I

1. A. S. Reber, "Implicit Learning of Artificial Grammars," *Journal of Verbal Learning and Verbal Behavior* 6 (1967): 855–63; L. R. Squire and S. Zola-Morgan, "Memory: Brain Systems and Behavior," *Trends in Neuroscience* 11 (1988): 170–75; K. S. Giovanello and M. Verfaellie, "Memory Systems of the Brain: A Cognitive Neuropsychological Analysis," *Seminars in Speech and Language* 22 (2001): 107–16.

2. S. Nicolas, "Experiments on Implicit Memory in a Korsakoff Patient by Claparède (1907)," *Cognitive Neuropsychology* 13 (1996): 1193–99.

3. B. Milner, "Memory Impairment Accompanying Bilateral Hippocampal Lesions," in *Psychologie De L'hippocampe*, eds, P. Passouant (Paris, France: Centre National de la Recherche Scientifique, 1962), 257-72.

4. Ibid.

5. S. Corkin, "Tactually-Guided Maze Learning in Man: Effects of Unilateral Cortical Excisions and Bilateral Hippocampal Lesions," *Neuropsychologia* 3 (1965): 339–51.

6. E. K. Miller and J. D. Cohen, "An Integrative Theory of Prefrontal Cortex Function," *Annual Review of Neuroscience* 24 (2001): 167–202; available online at web.mit.edu/ekmiller/Public/www/miller/Publications/Miller_Cohen _2001.pdf (accessed September 2012).

7. S. Corkin, "Acquisition of Motor Skill after Bilateral Medial Temporal Lobe Excision," *Neuropsychologia* 6 (1968): 255–65; available online at web.mit .edu/bnl/pdf/Corkin%201968.pdf (accessed September 2012).

8. Ibid.

9. Ibid.

10. Ibid.

11. Ibid.

12. Ibid.

13. Ibid.

14. G. Ryle, "Knowing How and Knowing That," in *The Concept of Mind* (London: Hutchinson's University Library, 1949), 26–60; full text available online at tinyurl.com/8kqedyj (accessed September 2012).

Decades after Ryle's book was published, the philosophical distinction between "knowing how" and "knowing that" made its way into the artificial intelligence community. As discussed at the beginning of Chapter 5, artificial-intelligence research has often helped to further theories about the brain because it deals with the practical task of programming computers to function like human brains. The resulting solutions can give neuroscientists models to test and predict how the brain works. In the 1970s, artificial-intelligence researchers used the terms *procedural* and *declarative* to describe two ways of representing knowledge. In

1975, Terry Winograd published an article, "Frame Representations and the Declarative/Procedural Controversy" (in *Representation and Understanding: Studies in Cognitive Sciences*, ed. D. G. Bobrow, et al. [New York: Academic Press], 185–210), which outlined an argument between the proceduralists and the declarativists: "The proceduralists assert that our knowledge is primarily a 'knowing how.' The human information processor is a stored program device, with its knowledge of the world *embedded* in the programs. What a person (or robot) knows about the English language, the game of chess, or the physical properties of his world is coextensive with his set of programs for operating with it" (p. 186). In other words, knowledge consists of the specific routines that guide our behavior. "The declarativists, on the other hand, do not believe that knowledge of a subject is intimately bound with the procedures for its use. They see intelligence as resting on two bases: a general set of procedures for manipulating facts of all sorts, and a set of specific facts describing particular knowledge domains." This way of viewing knowledge sees it as information rather than as a set of operations. Winograd advocated blurring the distinction between the two kinds of representations, and proposed taking the middle ground between declarative and procedural knowledge by specifying how particular declarative statements would be used. His idea was to attach procedures to facts in long-term memory.

In contrast, John Anderson argued for a fundamental difference between procedural and declarative knowledge. In his 1976 book *Language, Memory, and Thought* (Hillsdale, NJ: Psychology Press), with reference to Ryle, Anderson noted three distinguishing features. The first is that declarative knowledge is something we either have or lack, whereas procedural knowledge can be acquired gradually, a little at a time. A second distinction, he wrote, "is that one acquires declarative knowledge suddenly by being told whereas one acquires procedural knowledge gradually by performing the skill" (p. 117). The third distinctive feature is that we can tell someone about our declarative knowledge, but cannot explain our procedural knowledge.

While theorists debated how distinct these two kinds of knowledge really were, computer scientist Patrick Winston suggested a compromise. In his 1977 book *Artificial Intelligence* (Reading, MA: Addison-Wesley), Winston wrote, "There are arguments for and against the procedural and declarative positions on how knowledge should be stored. In most situations, the best plan is to face the problems in a bipartisan way drawing on talents from both sides of the aisle" (p. 393). Humans need procedural and declarative knowledge to function in everyday life, and the brain allocates different processes and circuits for obtaining and storing these two kinds of information. Milner had already revealed this biological distinction fifteen years earlier when she reported Henry's mirror-tracing results.

See also Milner, "Memory Disturbance after Bilateral Hippocampal Lesions."

15. M. Victor and A. H. Ropper, *Adams and Victor's Principles of Neurology*, 7th ed. (New York: McGraw-Hill, Medical Pub. Division, 2001).

16. Ibid.

17. Additional cognitive testing strengthened our conclusion that the mirror-tracing deficit we uncovered in Parkinson patients was truly a learning disorder. To rule out the possibility that the patients' slow learning was due to deficits in processing spatial layouts or basic motor functions, we asked them to perform additional tests to examine these other abilities. When our data analyses took all these other test scores into account, they still showed a significant learning deficit. This finding strengthens the view that mirror tracing is supported by a memory circuit that depends on intact neurotransmission in the striatum.

18. M. J. Nissen and P. Bullemer, "Attentional Requirements of Learning: Evidence from Performance Measures," *Cognitive Psychology* 19 (1987): 1–32.

19. D. Knopman and M. J. Nissen, "Procedural Learning Is Impaired in Huntington's Disease: Evidence from the Serial Reaction Time Task," *Neuropsychologia* 29 (1991): 245–54.

20. A. Pascual-Leone et al., "Procedural Learning in Parkinson's Disease and Cerebellar Degeneration," *Annals of Neurology* 34 (1993): 594–602; J. N. Sanes et al., "Motor Learning in Patients with Cerebellar Dysfunction," *Brain* 113 (1990): 103–20.

21. T. A. Martin et al., "Throwing while Looking through Prisms. I. Focal Olivocerebellar Lesions Impair Adaptation," and "II. Specificity and Storage of Multiple Gaze–Throw Calibrations," *Brain* 119 (1996): 1183–98, 1199–211.

22. R. Shadmehr and F. A. Mussa-Ivaldi, "Adaptive Representation of Dynamics during Learning of a Motor Task," *Journal of Neuroscience* 14 (1994): 3208–24; available online at www.jneurosci.org/content/14/5/3208.full.pdf+html (accessed September 2012).

One influential model from neuropsychology, which Daniel Willingham proposed in 1998, explains the stages through which motor-skill learning progresses. According to this theory, motor-skill learning engages two independent modes, one unconscious and the other conscious. The unconscious mode subsumes three motor control processes that function outside of awareness: selecting spatial targets for movement, sequencing these targets, and transforming them into muscle commands. The conscious, attention-demanding mode supports motor-skill learning by selecting goals to change the environment, selecting targets for movement, and assembling a sequence of targets. The conscious mode is exercised when a person imitates the performance of an expert. Learning advances through the interaction of the unconscious and conscious modes. Willingham's model allowed the researcher to make predictions about different learning stages and processes and their neural underpinnings. It did not, however, illuminate the mechanisms of how we learn motor skills step by step. See D. B. Willingham, "A Neuropsychological Theory of Motor Skill Learning," *Psychological Review* 105 (1998): 558–84.

23. M. Kawato and D. Wolpert, "Internal Models for Motor Control," *Novartis Foundation Symposium* 218 (1998): 291–304.

24. Ibid.

25. Ibid.

26. H. Imamizu and M. Kawato, "Brain Mechanisms for Predictive Control by Switching Internal Models: Implications for Higher-Order Cognitive Functions," *Psychological Research* 73 (2009): 527–44.

27. T. Brashers-Krug et al., "Consolidation in Human Motor Memory," *Nature* 382 (1996): 252–55; available online at tinyurl.com/8hhuga3 (accessed September 2012).

28. R. Shadmehr et al., "Time-Dependent Motor Memory Processes in Amnesic Subjects," *Journal of Neurophysiology* 80 (1998): 1590–97; available online at web.mit.edu/bnl/pdf/Shadmehr.pdf (accessed September 2012).

29. Ibid.

30. Ibid.

31. Ibid.

32. Ibid.

33. A. Karni et al., "The Acquisition of Skilled Motor Performance: Fast and Slow Experience-Driven Changes in Primary Motor Cortex," *Proceedings of the National Academy of Sciences* 95 (1998): 861–68.

34. Ibid. See also J. N. Sanes and J. P. Donoghue, "Plasticity and Primary Motor Cortex," *Annual Review of Neuroscience* 23 (2000): 393–415; available online at tinyurl.com/8oyl87x (accessed September 2012).

35. E. Dayan and L. G. Cohen, "Neuroplasticity Subserving Motor Skill Learning," *Neuron* 72 (2011): 443–54.

36. R. A. Poldrack et al., "The Neural Correlates of Motor Skill Automaticity," *Journal of Neuroscience* 25 (2005): 5356–64.

37. C. J. Steele and V. B. Penhune, "Specific Increases within Global Decreases: A Functional Magnetic Resonance Imaging Investigation of Five Days of Motor Sequence Learning," *Journal of Neuroscience* 30 (2010): 8332–41.

Chapter Nine: Memory without Remembering II

1. I. P. Pavlov, *Conditioned Reflexes: An Investigation of the Physiological Activity of the Cerebral Cortex* (London: Oxford University Press, 1927). Psychologist Edwin B. Twitmyer made a similar finding in humans nearly simultaneously. In 1902, he happened to observe that when a bell sounded just before a reflex hammer struck a person's knee and caused an involuntary kneejerk, the individual also exhibited this reflex on hearing the bell, even when the hammer did not

strike. Throughout the century following Pavlov and Twitmyer's discoveries, researchers examined classical conditioning in many species, including rats, crickets, fruit flies, fleas, and sea hares. E. B. Twitmyer, "Knee Jerks without Stimulation of the Patellar Tendon," *Psychological Bulletin* 2 (1905): 43–44; I. Gormezano et al., "Twenty Years of Classical Conditioning Research with the Rabbit," in *Progress in Physiological Psychology*, ed, J. M. Sprague et al. (New York: Academic Press, 1983), 197–275.

2. D. Woodruff-Pak, "Eyeblink Classical Conditioning in H.M.: Delay and Trace Paradigms," *Behavioral Neuroscience* 107 (1993): 911–25.

3. Ibid.

4. Ibid.

5. Ibid.

6. Ibid.

7. Ibid.

8. R. E. Clark et al., "Classical Conditioning, Awareness, and Brain Systems," *Trends in Cognitive Sciences* 6 (2002): 524–31.

9. Ibid.

10. Perceptual learning in amnesia was first reported in 1968 by neuropsychologists Elizabeth Warrington and Lawrence Weiskrantz. This discovery was almost as revolutionary as Milner's initial demonstration of Henry's preserved mirror-tracing skill. Five of their six amnesic patients had Korsakoff syndrome, in which cell loss occurs in the thalamus and hypothalamus, raising the question whether Henry's medial temporal-lobe lesions spared this ability. We eventually showed that he was indeed capable of perceptual learning. E. K. Warrington and L. Weiskrantz, "New Method of Testing Long-Term Retention with Special Reference to Amnesic Patients," *Nature* 217 (1968): 972–74.; B. Milner et al., "Further Analysis of the Hippocampal Amnesic Syndrome: 14-Year Follow-up Study of H.M.," *Neuropsychologia* 6 (1968): 215–34, available online at www.psychology.uiowa.edu/Faculty/Freeman/Milner_68.pdf (accessed November 2012).

11. E. S. Gollin," Developmental Studies of Visual Recognition of Incomplete Objects," *Perceptual and Motor Skills* 11 (1960): 289–98; Milner et al., "Further Analysis of the Hippocampal Amnesic Syndrome."

12. Milner et al., "Further Analysis of the Hippocampal Amnesic Syndrome."

13. Ibid.

14. Ibid.

15. J. Sergent et al., "Functional Neuroanatomy of Face and Object Processing. A Positron Emission Tomography Study," *Brain* 115 (1992): 15–36; N. Kanwisher, "Functional Specificity in the Human Brain: A Window into the Functional Architecture of the Mind," *Proceedings of the National Academy of Sciences* 107 (2010): 11163–70.

16. I. Gauthier et al., "Expertise for Cars and Birds Recruits Brain Areas Involved in Face Recognition," *Nature Neuroscience* 3 (2000): 191–97; available online at http://www.systems.neurosci.info/FMRI/gauthier00.pdf (accessed November 2012).

17. C. D. Smith et al., "MRI Diffusion Tensor Tracking of a New Amygdalo-Fusiform and Hippocampo-Fusiform Pathway System in Humans, *Journal of Magnetic Resonance Imaging* 29(2009): 1248–61.

18. Warrington and Weiskrantz published the first report of repetition priming in 1970, when they discovered that their amnesic patients could complete a three-letter stem—*MET*—to a previously studied word—*METAL*—just as often as control participants. The stimuli for this study were two fragmented versions of each word and the completed word. The investigators created the fragmented words by photographing them with patches covering parts of each letter. In an initial study phase, participants first saw the most-fragmented version of all the words, next the less-fragmented version of each, and then the complete word—*METAL*. They were asked to identify the word as quickly as possible. The purpose of the study was to compare three measures of memory retention: two of them—recall and recognition—were declarative, and the third—partial completion—was nondeclarative. The amnesic group, as expected, had trouble recalling and recognizing the words they had seen before—the hallmark of amnesia. The big shock came in the subsequent test phase, when participants viewed the first three letters of each word and then thought of a five-letter word—retrieval by partial completion. On this measure, the amnesic patients reported as many studied words as the control participants.

Although the researchers did not at the time interpret this result as evidence of spared priming in amnesia, they still published the first demonstration of word-stem completion priming. The method they used gave scientists a way to explore learning without awareness in healthy individuals and in patients with a variety of neurological and psychiatric disorders. See E. K. Warrington and L. Weiskrantz, "Amnesic Syndrome: Consolidation or Retrieval?," *Nature* 228 (1970): 628–30.

During the 1980s and '90s, hundreds of research reports on repetition priming appeared. Memory researchers examined priming effects in healthy participants and amnesic patients, using a wide assortment of text stimuli: words, pseudowords (made-up words that obey the rules of English orthography), word fragments, categories of objects, homophones (words that sound the same but have different meanings, such as *bear* and *bare*), pictures, fragmented pictures, and patterns. These elegant studies elucidated the cognitive intricacies of the priming effect, particularly in healthy young adults.

As the body of knowledge on repetition priming grew, the question of preserved repetition-priming in amnesia continued to engage memory experts. In

1984, Peter Graf, Larry Squire, and George Mandler wrote a high-profile paper in which they reported the results of three experiments. Their intent was to compare the performance of amnesic and control participants on four measures of learning: three of them—free recall, recognition, and cued recall—were declarative, whereas the fourth—word completion—was nondeclarative. Participants initially studied a list of words and then took one of the four tests just mentioned. P. Graf et al., "The Information That Amnesic Patients Do Not Forget," *Journal of Experimental Psychology: Learning, Memory, and Cognition* 10 (1984): 164–78.

For the free-recall test, participants wrote the words they could remember from the study list on a sheet of paper. For the recognition test, they saw one of the studied words with two others that began with the same three-letter stem. When the studied word was *MARket*, the distractor words were *MARy* and *MARble*. Participants had to pick out the word they had seen previously. For the cued-recall and word-completion tests, participants received the first three letters of the studied words as cues. The critical difference between these two tasks was in the instructions. For cued recall, participants were asked to recall intentionally the list of words with the help of the cues. It was clear to all participants that that this was a memory test. For word completion, participants were told that the three-letter stem was the beginning of an English word, and were asked to make each into a word. They were encouraged to write the first word that came to mind, and were unaware that their memory was being tested.

The findings validated the 1970 results of Warrington and Weiskrantz. The free-recall, recognition, and cued-recall tasks all measured declarative memory, and performance on those tasks, not surprisingly, was severely impaired in the amnesic participants. The key issue was whether amnesic patients would perform the word-completion task like controls—and they did. With reference to the explanation of priming as activation of an established representation of the studied words, the scientists concluded that this kind of activation was intact in amnesia. This experiment illustrated the crucial role of instructions in the distinction between declarative ad nondeclarative memory. When the amnesic patients were explicitly asked to recall the list of words with the help of the three-letter stem, they had to access their declarative knowledge, and therefore failed to perform comparably to healthy participants. When they were allowed to rely on their nondeclarative knowledge, however, and simply complete the three-letter stem with the first word that leaped to mind, they could accomplish the task just as successfully as the controls. See Warrington and Weiskrantz, "Amnesic Syndrome: Consolidation or Retrieval?"; R. Diamond and P. Rozin, "Activation of Existing Memories in Anterograde Amnesia," *Journal of Abnormal Psychology* 93 (1984): 98–105, available online at http://www.psych.stanford.edu/~jlm/pdfs/DiamondRozin84.pdf (accessed November 2012).

These findings raised a pivotal question: Does the priming effect last as long in amnesic patients as it does in controls? For the amnesic patients' performance to be considered normal, the answer to this question would have to be yes. Examiners showed participants the study list and then tested them either immediately, after fifteen minutes, or after 120 minutes, using a different set of words at each delay. The patients achieved as many correct responses as controls, and the performance of the two groups was comparable across the three delays. This result meant that the priming effect lasted the same amount of time—two hours—in amnesic patients and in controls. See Warrington and Weiskrantz, "Amnesic Syndrome: Consolidation or Retrieval?"; Graf et al., "The Information That Amnesic Patients Do Not Forget."

19. J.D.E. Gabrieli et al., "Dissociation among Structural-Perceptual, Lexical-Semantic, and Event-Fact Memory Systems in Amnesia, Alzheimer's Disease, and Normal Subjects," *Cortex* 30 (1994): 75–103.

20. Ibid.

21. Ibid.

22. Diamond and Rozin, "Activation of Existing Memories in Anterograde Amnesia."

23. Ibid.

24. J.D.E. Gabrieli et al., "Intact Priming of Patterns Despite Impaired Memory," Neuropsychologia 28 (1990): 417–27; available online at http://web.mit.edu/bnl/pdf/Gabrieli_Milberg_Keane_Corkin_1990.pdf (accessed November 2012).

25. Ibid.

26. Ibid.

27. Ibid.

28. Ibid.

29. M. M. Keane et al., "Priming in Perceptual Identification of Pseudowords Is Normal in Alzheimer's Disease," *Neuropsychologia* 32 (1994): 343–56.

30. Keane et al., "Priming of Perceptual Identification of Pseudowords Is Normal in Alzheimer's Disease"; M. M. Keane et al., "Evidence for a Dissociation between Perceptual and Conceptual Priming in Alzheimer's Disease," *Behavioral Neuroscience* 105 (1991): 326–42.

31. Ibid.

32. S. E. Arnold et al., "The Topographical and Neuroanatomical Distribution of Neurofibrillary Tangles and Neuritic Plaques in the Cerebral Cortex of Patients with Alzheimer's Disease," *Cerebral Cortex* 1 (1991): 103–16.

33. M. M. Keane et al., "Double Dissociation of Memory Capacities after Bilateral Occipital-Lobe or Medial Temporal-Lobe Lesions," *Brain* 118 (1995): 1129–48.

Chapter Ten: Henry's Universe

1. J. A. Ogden and S. Corkin, "Memories of H.M.," in *Memory Mechanisms: A Tribute to G. V. Goddard*, ed. M. Corballis et al. (Hillsdale, NJ: L. Erlbaum Associates, 1991), 195–215.

2. N. Hebben et al., "Diminished Ability to Interpret and Report Internal States after Bilateral Medial Temporal Resection: Case H.M.," *Behavioral Neuroscience* 99 (1985): 1031–39; available online at web.mit.edu/bnl/pdf/Diminished %20Ability.pdf (accessed November 2012).

3. S. Kobayashi, "Organization of Neural Systems for Aversive Information Processing: Pain, Error, and Punishment," *Frontiers in Neuroscience* 6 (2012); available online at www.ncbi.nlm.nih.gov/pmc/articles/PMC3448295/ (accessed November 2012).

4. Hebben et al., "Diminished Ability"; W. C. Clark, "Pain Sensitivity and the Report of Pain: An Introduction to Sensory Decision Theory," *Anesthesiology* 40 (1974): 272–87.

5. Hebben et al., "Diminished Ability"; C. de Graaf et al., "Biomarkers of Satiation and Satiety," *American Journal of Clinical Nutrition* 79 (2004): 946–61.

6. Hebben et al., "Diminished Ability."

7. N. Butters and L. S. Cermak, "A Case Study of the Forgetting of Autobiographical Knowledge: Implications for the Study of Retrograde Amnesia," In *Autobiographical Memory*, ed. D. C. Rubin (New York: Cambridge University Press, 1986), 253–72.

8. W. B. Scoville and B. Milner, "Loss of Recent Memory after Bilateral Hippocampal Lesions," *Journal of Neurology, Neurosurgery, and Psychiatry* 20 (1957): 11–21; B. Milner et al., "Further Analysis of the Hippocampal Amnesic Syndrome: 14-Year Follow-up Study of H.M.," *Neuropsychologia* 6 (1968): 215–34, available online at www.psychology.uiowa.edu/Faculty/Freeman/Milner _68.pdf (accessed November 2012).

9. H. J. Sagar et al., "Dissociations among Processes in Remote Memory," *Annals of the New York Academy of Sciences* 444 (1985): 533–55.

10. Ibid.

11. Sagar et al., "Dissociations among Processes"; H. F. Crovitz and H. Schiffman, "Frequency of Episodic Memories as a Function of Their Age," *Bulletin of the Psychonomic Society* 4 (1974): 517–18.

12. Sagar et al., "Dissociations among Processes"; H. J. Sagar et al., "Temporal Ordering and Short-Term Memory Deficits in Parkinson's Disease," *Brain* 111 (Pt 3) (1988): 525–39. The "last in, first out" theory dates from 1881, when French psychologist Théodule Ribot noted that retrograde memory loss often follows a temporal gradient—newer memories are more likely to be lost,

while older memories are more likely to be preserved. T. Ribot, *Les Maladies de la Mémoire* (Paris: Germer Baillière, 1881).

13. The results of our structured interview with Henry showed that the early clinical reports from the 1950s and 60s had grossly underestimated the length of his retrograde amnesia. The processes needed to retrieve memories from the years before his operation were severely disrupted. One of our tasks, the Crovitz test, was later criticized by us and others on the grounds that it was not sufficiently sensitive in differentiating between memory reports that were just detailed enough to merit the maximum score of three and others that also earned a three but were much richer in detail. We needed a test that would make a finer distinction among the top performers. Our subsequent experiments achieved that goal. Crovitz and Schiffman, "Frequency of Episodic Memories."

14. Taken together, our early studies of Henry's retrograde amnesia yielded conflicting results, in part, because they did not distinguish between personal remote memories that reflected general knowledge, such as the name of a high school, and those that relied on reliving an experience, such as a first kiss. As we saw from Henry's case, he could tell us what high school he had attended, but not what happened on his graduation day. When we questioned him, it was apparent that although he could provide answers to the most general questions, something was missing when we pressed him for more specifics.

15. E. Tulving, "Episodic and Semantic Memory," in *Organization of Memory*, ed. E. Tulving and W. Donaldson (New York: Academic Press, 1972), 381–403.

16. L. R. Squire, "Memory and the Hippocampus: A Synthesis from Findings with Rats, Monkeys, and Humans," *Psychological Review* 99 (1992): 195–231.

17. L. R. Squire and P. J. Bayley, "The Neuroscience of Remote Memory," *Current Opinion in Neurobiology* 17 (2007): 185–96.

18. L. Nadel and M. Moscovitch, "Memory Consolidation, Retrograde Amnesia and the Hippocampal Complex," *Current Opinion Neurobiology* 7 (1997): 217–27; also Moscovitch and Nadel, "Consolidation and the Hippocampal Complex Revisited: In Defense of the Multiple-Trace Model," *Current Opinion Neurobiology* 8 (1998): 297–300.

19. B. Milner, "The Memory Defect in Bilateral Hippocampal Lesions," *Psychiatric Research Reports of the American Psychiatric Association* 11 (1959): 43–58.

20. S. Steinvorth et al., "Medial Temporal Lobe Structures Are Needed to Re-experience Remote Autobiographical Memories: Evidence from H.M. and W.R.," *Neuropsychologia* 43 (2005): 479–96.

21. Ibid.; See also E. A. Kensinger and S. Corkin, "Two Routes to Emotional Memory: Distinct Neural Processes for Valence and Arousal," *Proceedings of*

the National Academy of Sciences 101 (2004): 3310–15; available online at www.pnas.org/content/101/9/3310.full.pdf+html (accessed November 2012).

22. Steinvorth et al., "Medial Temporal Lobe Structures."

23. L. R. Squire, "The Legacy of Patient H.M. for Neuroscience," *Neuron* 61 (2009): 6–9; available online at whoville.ucsd.edu/PDFs/444_Squire _Neuron_2009.pdf (accessed November 2012); S. Corkin et al., "H.M.'s Medial Temporal Lobe Lesion: Findings from MRI," *Journal of Neuroscience* 17 (1997): 3964–79.

24. Steinvorth et al., "Medial Temporal Lobe Structures Are Needed"; Nadel and Moscovitch, "Memory Consolidation, Retrograde Amnesia, and the Hippocampal Complex."

25. Y. Nir and G. Tononi, "Dreaming and the Brain: From Phenomenology to Neurophysiology," *Trends in Cognitive Sciences* 14 (2010): 88–100.

26. P. Maquet et al., "Functional Neuroanatomy of Human Rapid-Eye-Movement Sleep and Dreaming," *Nature* 383 (1996): 163–66.

27. D. L. Schacter et al., "Episodic Simulation of Future Events: Concepts, Data, and Applications," *Annals of the New York Academy of Sciences* 1124 (2008): 39–60.

Chapter Eleven: Knowing Facts

1. E. A. Kensinger et al., "Bilateral Medial Temporal Lobe Damage Does Not Affect Lexical or Grammatical Processing: Evidence from Amnesic Patient H.M.," *Hippocampus* 11 (2001): 347–60.

2. J. R. Lackner, "Observations on the Speech Processing Capabilities of an Amnesic Patient: Several Aspects of H.M.'s Language Function," *Neuropsychologia* 12 (1974): 199–207.

3. D. G. MacKay et al., "H.M. Revisited: Relations between Language Comprehension, Memory, and the Hippocampus System," *Journal of Cognitive Neuroscience* 10 (1998): 377–94.

4. Kensinger et al., "Bilateral Medial Temporal Lobe Damage Does Not Affect Lexical or Grammatical Processing."

5. Ibid.

6. Ibid.

7. A. D. Friederici, "The Brain Basis of Language Processing: From Structure to Function," *Physiological Review* 92 (2011): 1357–92; C. J. Price, "A Review and Synthesis of the First 20 Years of PET and fMRI Studies of Heard Speech, Spoken Language, and reading," *Neuroimage* 62 (2012): 816–47.

8. D. C. Park and P. Reuter-Lorenz, "The Adaptive Brain: Aging and Neurocognitive Scaffolding," *Annual Review of Psychology* 60 (2009): 173–96.

9. Kensinger et al., "Bilateral Medial Temporal Lobe Damage Does Not Affect Lexical or Grammatical Processing."

10. E. K. Warrington and L. Weiskrantz, "Amnesic Syndrome: Consolidation or Retrieval?," *Nature* 228 (1970): 628–30; W. D. Marslen-Wilson and H.-L. Teuber, "Memory for Remote Events in Anterograde Amnesia: Recognition of Public Figures from Newsphotographs," *Neuropsychologia* 13 (1975): 353–64.

11. M. Kinsbourne and F. Wood, "Short-Term Memory Processes and the Amnesic Syndrome," in *Short-Term Memory*, eds, D. Deutsch et al. (San Diego, CA: Academic Press, 1975), 258–93; M. Kinsbourne, "Brain Mechanisms and Memory," *Human Neurobiology* 6 (1987): 81–92.

12. J. D. Gabrieli et al., "The Impaired Learning of Semantic Knowledge Following Bilateral Medial Temporal-Lobe Resection," *Brain Cognition* 7 (1988): 157–77.

13. Ibid.

14. F. B. Wood et al., "The Episodic-Semantic Memory Distinction in Memory and Amnesia: Clinical and Experimental Observations," in *Human Memory and Amnesia*, eds, L. S. Cermak (Hillsdale, NJ: Erlbaum, 1982), 167–94.

15. J. D. Gabrieli et al., "The Impaired Learning of Semantic Knowledge."

16. Ibid.

17. Ibid.

18. Ibid.

19. B. R. Postle and S. Corkin, "Impaired Word-Stem Completion Priming but Intact Perceptual Identification Priming with Novel Words: Evidence from the Amnesic Patient H.M.," *Neuropsychologia* 36 (1998): 421–40.

20. Ibid.

21. Ibid.

22. Ibid.

23. Ibid.

24. Ibid.

25. Ibid.

26. E. Tulving et al., "Long-Lasting Perceptual Priming and Semantic Learning in Amnesia: A Case Experiment," *Journal of Experimental Psychology: Human Learning and Memory* 17 (1991): 595–617; P. J. Bayley and L. R. Squire, "Medial Temporal Lobe Amnesia: Gradual Acquisition of Factual Information by Nondeclarative Memory," *Journal of Neuroscience* 22 (2002): 5741-8.

27. G. O'Kane et al., "Evidence for Semantic Learning in Profound Amnesia: An Investigation with Patient H.M.," *Hippocampus* 14 (2004); 417–25.

28. Ibid.

29. Ibid.

30. Ibid.

31. Ibid.

32. Ibid.

33. Ibid.

34. Ibid.

35. Ibid.

36. B. G. Skotko et al., ""Puzzling Thoughts for H.M.: Can New Semantic Information Be Anchored to Old Semantic Memories?," *Neuropsychology* 18 (2004): 756–69.

37. Ibid.

38. Ibid.

39. F. C. Bartlett, *Remembering: A Study in Experimental and Social Psychology* (Cambridge: University Press, 1932).

40. D. Tse et al., "Schemas and Memory Consolidation," *Science* 316 (2007): 76–82.

41. Ibid.

Chapter Twelve: Rising Fame and Declining Health

1. W. B. Scoville and B. Milner, "Loss of Recent Memory after Bilateral Hippocampal Lesions," *Journal of Neurology, Neurosurgery, and Psychiatry* 20 (1957): 11–21.

2. D. H. Salat et al., "Neuroimaging H.M.: A 10-Year Follow-up Examination," *Hippocampus* 16 (2006): 936–45.

3. Our brains house billions of individual nerve cells or neurons, with thousands of distinct types already identified and others still unknown. Neurons are dedicated to information processing—receiving, conducting, and transmitting electrical and chemical signals. The standard neuron has a nerve-cell body, numerous dendrites, and a single axon that branches out. The dendrites receive signals from other cells and deliver them to the cell body, while the axon carries signals away from the cell body to activate other neurons. Clusters of nerve-cell bodies are called gray matter, and collections of axons are referred to as white matter. The cerebral cortex is made up of gray matter, while the short and long fiber pathways that allow information to flow from one area to another are white matter. With MRI, we have been able to observe the effects of brain aging in both gray matter and white matter in healthy older people and in Henry (E. Diaz, "A Functional Genomics Guide to the Galaxy of Neuronal Cell Types," *Nature Neuroscience* 9 (2006): 10–12; K. Sugino et al., "Molecular Taxonomy of Major Neuronal Classes in the Adult Mouse Forebrain," *Nature Neuroscience* 9 (2006): 99–107).

One autopsy study of older adults who were not demented found that the gray matter in the cortex became much thinner with increasing age, but cortical thickness did not correlate with performance on a cognitive test that estimated the individual's overall mental capacity just before death. This finding suggested that cognition in aging might be more closely linked to the loss of white matter than gray matter. If that were the case, then what would be the consequence of white-matter breakdown? When the white matter is intact, the transmission of neural information is fast and on course, a river flowing swiftly with no obstacles in the way. But when minute structures in the white matter are compromised, this river is cluttered with dams, rocks, trees, and a partially submerged boat. Neural transmission becomes obstructed and inefficient, slowing neural and cognitive processing. (S. H. Freeman et al., "Preservation of Neuronal Number Despite Age-Related Cortical Brain Atrophy in Elderly Subjects without Alzheimer Disease," *Journal of Neuropathology and Experimental* Neurology 67 (2008): 1205–12; T. A. Salthouse, "The Processing-Speed Theory of Adult Age Differences in Cognition," *Psychological Review* 103 (1996): 403–28).

Although white-matter changes are more conspicuous than gray-matter changes in older brains, mapping out these networks in the living brain was difficult until recently. An advanced type of MRI, diffusion-tensor imaging, can now measure and map white-matter tissue integrity in living individuals in health and disease. With this tool, our collaborators at Mass General Martinos Center found erosion of white matter not only in older participants but also in middle-aged individuals, reinforcing that age-related decline in memory may also begin in middle age, even in perfectly healthy people.

In 2008, my lab members and I asked two key questions: are white matter and gray matter affected differently by aging, and are measures of cognitive performance more closely linked to alterations in white matter or to alterations in gray matter? We used advanced MRI techniques to measure gray matter thickness and subtle changes in white matter across the entire brain in young and older participants. The imaging data provided information about the integrity of brain areas that mediated three kinds of abilities: episodic memory, delayed recall of word lists and stories; semantic memory, naming objects and vocabulary; and cognitive control processes—paying attention, overriding dominant responses, and achieving goals. On the cognitive tests, the young adults outperformed the older adults on measures of episodic memory and cognitive control, but, as is often the case, the older adults performed better than the young on the semantic memory tasks, which probed their general knowledge about the world. As we age, our vocabulary and pool of information grow and become more sophisticated.

Consistent with previous studies, we found that healthy aging is accompanied by deterioration of gray and white matter. A further analysis of the MRI images revealed new insights. When we correlated these measures of brain structure with the older adults' cognitive-test scores, we found that cortical thickness—the indicator of gray matter—was unrelated to performance on the cognitive tests. Instead, our results confirmed the speculation that white-matter damage is largely responsible for the cognitive deficits that characterize healthy aging. We found region-specific correlations between cognitive-test scores and measures of white matter. Cognitive-control processes correlated with the integrity of frontal-lobe white matter, whereas episodic memory was related to the integrity of temporal- and parietal-lobe white matter. Our experiment sent the important message that scientists who want to understand the neural underpinnings of cognitive loss should examine not just gray matter, but white matter as well. This suggestion applies not just to experiments on aging and diseases of aging, but also to research on participants of all ages. (D. A. Ziegler et al., "Cognition in Healthy Aging Is Related to Regional White Matter Integrity, but Not Cortical Thickness," *Neurobiology of Aging* 31 (2010): 1912–26); D. H. Salat et al., "Age-Related Alterations in White Matter Microstructure Measured by Diffusion Tensor Imaging," *Neurobiology of Aging* 26 (2005): 1215–27.

4. J. W. Rowe and R. L. Kahn, "Human Aging: Usual and Successful," *Science* 237 (1987): 143–9.

5. Salat et al., "Neuroimaging H.M."

6. Ibid.

Chapter Thirteen: Henry's Legacy

1. I had long believed that it would be essential to study Henry's brain after his death to take full advantage of the wealth of information that evolved from his research participation. My view about brain donation stemmed partly from recognizing the value of information that had been gained from previous autopsy studies in Parkinson and Alzheimer disease. In 1960, a neuroscientist at the University of Vienna conducted autopsies of patients with Parkinson disease, and found that they had lower than normal levels of dopamine in their brain. This major discovery spurred treatments to replace dopamine function and thereby relieve the abnormal movements that characterize Parkinson disease.

In Alzheimer disease, we can know for sure that a patient had the disease only by looking for pathological markers in the brain at autopsy. Even in the early stages of the disorder, neurofibrillary tangles and amyloid plaques are abundant, and cell death is significant.

Examining brains after death also enlightens scientists who study cognitive functions in other manifestations of brain damage, although autopsies on such cases are rare. In this book, I have emphasized how much we have learned about the roles of different brain circuits by studying patients who had the misfortune of losing function in those areas; Henry was just one outstanding example. In most of our research, we make guesses about the actual brain damage. When I studied brain injuries in military veterans, for instance, I had to infer the location and extent of their brain lesions based on the wounds in their skulls. Recent imaging advances have made it possible to see brain anatomy in much more detail, but MRI is still imperfect. The only way to truly see brain abnormalities is to look directly at the brain, which is possible only after death. Postmortem studies of these patients tell us, in a more detailed and complete way, what brain damage led to their cognitive deficits, and they could inform scientific debates about the role of specific brain structures in memory and other capacities.

2. "H.M., an Unforgettable Amnesiac, Dies at 82"; www.nytimes.com /2008/12/05/us/05hm.html?pagewanted=all (accessed December 2012).

Epilogue

1. Experimental surgeries have been practiced since antiquity and have been the bedrock of many treatment advances. A startling twenty-first-century example is natural-orifice surgery. Several years ago, a Mass General surgeon extracted a woman's gallbladder via her vagina. This kind of surgery, unpleasant as it sounds, offers several advantages over time-honored methods. These procedures do not require an incision because they are carried out through a natural opening in the body—the mouth, anus, vagina, or urethra. As a result, they do not leave a scar, and the recovery time is much faster—days instead of weeks, in the case of gallbladder removal. Although natural-orifice surgery seems safe and effective, we cannot draw firm conclusions about the experiments until each procedure and its specialized new tools have been tested in clinical trials. In contrast, Henry's experimental operation immediately gave a strong directive to other surgeons—*never perform this operation.* Sacha Pfeiffer, "You Want to Take My What Out of My Where? Hospitals Experiment with Orifice Surgery," WBUR / NPR News, June 22, 2009, www.wbur.org/2009 /06/22/orifice-surgery (accessed December 2012).

2. B. Milner and W. Penfield, "The Effect of Hippocampal Lesions on Recent Memory," *Transactions of the American Neurological Association* (1955–56): 42–48; W. B. Scoville, "World Neurosurgery: A Personal History of a Surgical Specialty," *International Surgery* 58 (1973): 526–35.

3. S. Tigaran et al., "Evidence of Cardiac Ischemia during Seizures in Drug Refractory Epilepsy Patients," *Neurology* 60 (2003): 492–95.

Index